TM & © 1994-2000. The Strategic Coach Inc. All rights reserved. The Strategic Coach®, Strategic Coach®, The Strategic Coach Program™, Great Crossover®, Free Days™, Focus Days™, Buffer Days ™, and The Entrepreneurial Time System™ are trademarks of The Strategic Coach Inc.

No part of this publication may be reproduced in any form, or by any means whatsoever, without written permission from the publisher, except in the case of brief quotations embodied in critical articles and reviews.

Printed in Toronto, Canada. May 2000. The Strategic Coach Inc., 33 Fraser Avenue, Suite 201, Toronto, Ontario, M6K 3J9.

This publication is meant to strengthen your common sense, not to substitute for it. It is also not a substitute for the advice of your doctor, lawyer, accountant, or any of your advisors, personal or professional.

SECOND EDITION, Paperback version

Canadian Cataloguing in Publication Data

Sullivan, Dan, 1944-
The great crossover: personal confidence in the age of the microchip

Rev. ed
ISBN 1-896635-13-X

1. Success in business. 2. Entrepreneurship 3. I. Smith, Babs, 1951- II. Néray, Michel, 1957- III. Title

HB615.S94 2000   650.1   C99-931331-2

The Great Crossover

Personal Confidence In The Age Of The Microchip

table of contents

introduction	The Great Crossover	1
strategy 1	Make The Two Entrepreneurial Decisions And Never Look Back	9
strategy 2	Get Away From Bureaucracies As Far As You Can	15
strategy 3	Obey The Laws Of The Microchip	21
strategy 4	Let The New Technology Come To You	31
strategy 5	Give Children An Entrepreneurial Attitude	37
strategy 6	Decide To Be Happy Right Now And Live With It	43
strategy 7	Choose Creativity And Stop Complaining	47
strategy 8	Focus On Progress And Forget About Perfection	53
strategy 9	Focus On Habits And Forget About Discipline	59
strategy 10	Focus On Your Strengths And Delegate Your Weaknesses	65

strategy 11	Improve Your Mastery Of The English Language	**71**
strategy 12	Buy Peace Of Mind In A Debtor's World With 20% Savings	**79**
strategy 13	Eat, Sleep, Exercise, And Meditate	**87**
strategy 14	Increase Your Free Days To Multiply Productivity	**93**
strategy 15	Visit Mountains, Forests, Deserts, and Oceans	**99**
strategy 16	Decide How Long You're Going To Live, and Plan Backwards	**105**
strategy 17	Plan Your Life In Three-Year Quantum Leaps	**111**
strategy18	Team Up and Prosper With Female Entrepreneurs	**119**
strategy 19	Ask Everybody You Meet The Three-Year Question	**125**
strategy 20	Grow To The Genius Level Of Personal Confidence	**131**
background	The Strategic Coach Program	**140**

introduction

The Great

Crossover

Since the late 1980s, there has been a widespread awareness developing that a new force — microtechnology — is drastically changing the way human beings live on the planet.

People are aware that something big is happening, but they're not sure how to think about it, or what to do about it. Throughout every society, community, organization, and group, there is a sense of unease about a technological future that undermines personal confidence.

In fact, we have entered a period of global transformation that rivals at least three other extraordinary periods in history. In each case, the development of a new form of language led to a breakthrough in the ability of humankind to communicate, share knowledge, and live and work together. This in turn caused the total upheaval of all the concepts, values, methods, structures, systems, and balances of power and wealth that defined the status quo.

The three forms of language which brought about the previous "great crossovers" were spoken language, written language, and printed language. Now, the utilization of the microchip — and the invention of digital language — is bringing about the fourth great crossover, and all of civilization up to this point is being profoundly transformed in the process. At the personal level, this new way of communicating is fundamentally transforming how individuals in every society and culture organize and conduct their lives. As a result, all of planetary society is making a crossover to a new kind of civilization based on creativity, cooperation, and competition made possible by microtechnology.

Many people feel that this means we must become technologically literate, that we must learn how to master sophisticated technological devices and systems. On every side, there are encouragements (or warnings) to become part of "the information age," to plug into "the electronic future," or to drive onto "the electronic superhighway." Never has this been so evident as at the beginning of the new millennium with the excitement surrounding the "dot-com" revolution.

But even those who have mastered the machines are still feeling anxious.

Those whose jobs are based on microtechnology — those who are on the cutting edge of new methods and applications — seem to be suffering from the same personal insecurities and problems as those who are technologically ignorant. They may be more skillful with computer technology, but they aren't necessarily more confident about where their lives are going. In fact, for many, the deeper they get into this unpredictable universe of microtechnology, the more anxiety-filled their lives seem to become. In other words, technological knowledge and skill in themselves don't provide us with personal confidence in the age of the microchip. It was with this in mind that this book was written.

This book is not about microchips or computers, but about the personal attitudes and habits needed to

live creatively and with confidence in a world governed by the microchip.

Although we are still in the early stages of this fourth great crossover — with the enormous turmoil and disruption that still lie ahead — it is already possible to see the two great historical trends of the 21st century:

The collapse of bureaucracy and the rise of entrepreneurism.

The increased use of microtechnology works against the continued existence of rigid, hierarchical organizations, and it works for the unlimited exercise of individual creativity and enterprise. As these trends gain momentum, those individuals who have depended on bureaucratic employment and position will experience a loss of their security and status, while those who create their own enterprises will assume greater power and influence. This has enormous implications for every aspect of human life, most of which cannot be accurately predicted yet.

But already, we can see that in modern, industrialized countries, the division between the "haves" and "have-nots" is becoming more pronounced. The haves are those who understand the entrepreneurial opportunities offered by microtechnology; the have-nots are those who do not yet understand how microtechnology works in their lives and who feel victimized by the confusing changes.

The trends of the great crossover

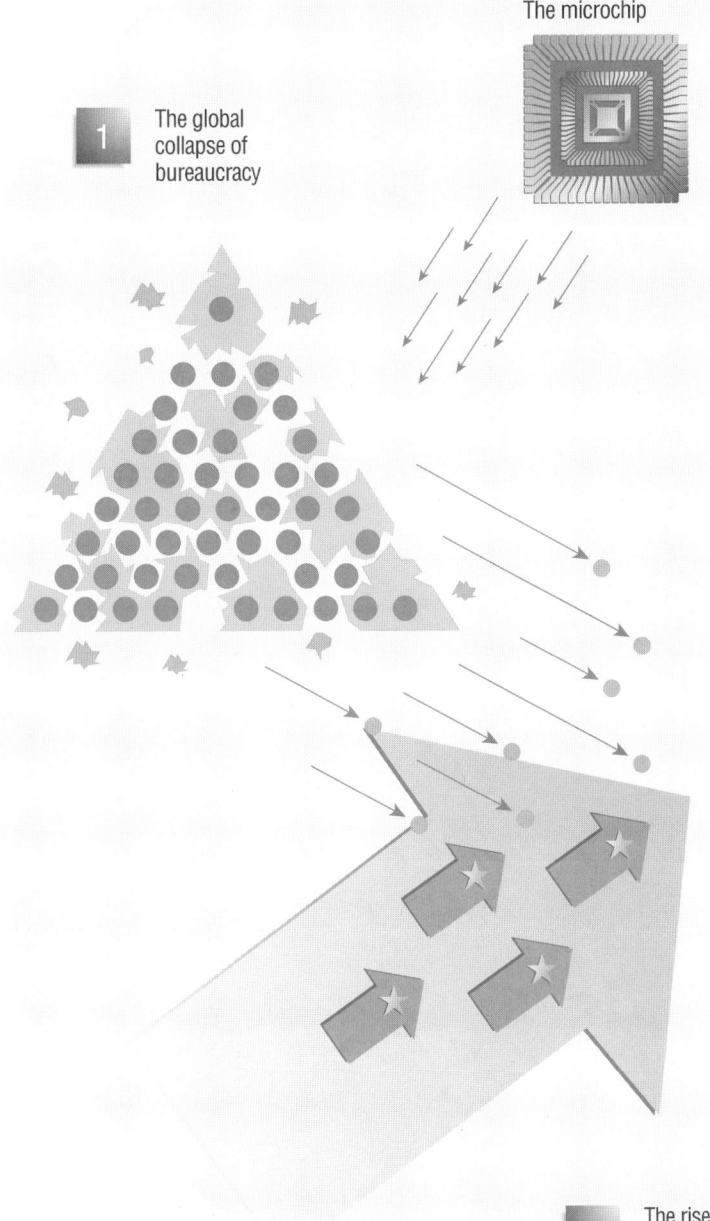

The microchip

1. The global collapse of bureaucracy

2. The rise of entrepreneurism

The educational system as a whole — now heavily bureaucratic and unionized — is only beginning to grasp the significance of the great crossover from a world based on printed language to one where family life, work, religion, art, culture, politics, and the most productive economic activities on the planet will be conducted electronically. But the deficiencies of the educational system do not prevent a great many people from teaching themselves and their children.

Millions of individual men and women are developing personal strategies for this new microchip world — not through formal instruction, but through economic risk-taking.

These individuals, who earn their living as entrepreneurs, have learned through practical failure and success how to maintain a sense of perspective, focus, creativity, and balance during times of enormous change. As entrepreneurs, they have learned how to thrive on change by becoming agents of change.

As we begin the 21st Century, there is not yet an overall philosophy based on microtechnology. And for the next 50 years, there probably still will be no overall understanding of what is taking place in the world. The changes caused by microtechnology are too massive, too profound, too far-reaching, and are occurring much too quickly for any one individual or group of individuals to comprehend and articulate. And while countless articles, books, speeches, and programs have already been devoted to the impact of the microchip on every area of human activity, there is still no universally accepted strategy or framework we can use to help us through these tumultuous times. Yet, in the absence of "One Big Wisdom," there will be many small wisdoms that complement and support each other.

During this next half-century, the most useful knowledge will be found on the individual level rather than the institutional level.

The role models that our children, grandchildren, and great-grandchildren will emulate over these next three generations will not be found among the traditional authority figures of society, but rather among the inventors of new products, services, businesses, business models, and industries.

The personal strategies that follow are about microtechnology, but not from a technological perspective. They are not about technically mastering computers or any other kind of equipment. Rather, these strategies are about mastering one's overall life in a global economy governed by the microchip, through the development of an entrepreneurial approach to all aspects of personal and professional existence over an entire lifetime. As more and more individuals adapt these and other entrepreneurial strategies to their unique circumstances over the next 50 years, a collective new wisdom will gradually emerge in all parts of the planet. It will be the basis for a global entrepreneurial society that will develop and prosper for hundreds of years into the future.

strategy 1

Make The Two Entrepreneurial Decisions And Never Look Back

Periods of great change have always favored the entrepreneur. Wars, natural disasters, and great inventions create new needs and upset existing balances in society. The entrepreneur is the individual who takes personal advantage of the changes that confuse and paralyze others.

Large organizations — which profit by maintaining the status quo, based on mass production and marketing — are generally slow to react and adapt. They leave plenty of room for quick and agile entrepreneurs to fill the gaps in the market, almost like a natural force that works to re-establish equilibrium in the environment. Then, as society returns to a new equilibrium, new empires grow from the seeds that were sown by entrepreneurs.

Today, there are more individuals becoming entrepreneurs than during any other period in history, for reasons that go well beyond the simple desire to profit from change.

People are becoming entrepreneurs out of necessity.

It is increasingly clear that bureaucracies are no longer able to offer the financial security they once promised. Employees have little or no control over their livelihood, their professional objectives, or even their day-to-day activities. Governments and large corporations have become unnerving places in which to work. Even the most bureaucratic staff member or government functionary is beginning to concede that the entrepreneur actually faces fewer risks and has greater control over his or her future.

The two entrepreneurial commitments

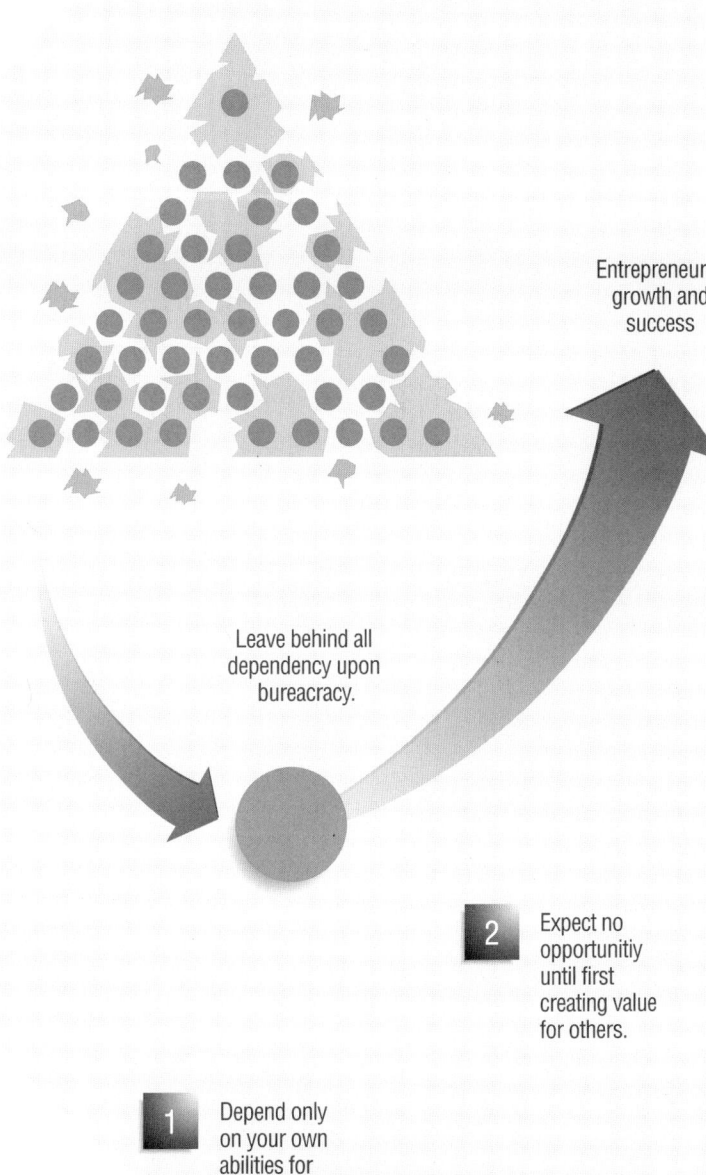

Entrepreneurial growth and success

Leave behind all dependency upon bureacracy.

2 Expect no opportunitiy until first creating value for others.

1 Depend only on your own abilities for financial security.

This historically unique situation can be traced back to microchip technologies. With their incredible ability to perform certain kinds of work, computers are one of the most destabilizing forces in history, while at the same time they are the most empowering tool ever developed.

Microtechnology allows individuals to leverage their knowledge into successful businesses, with a minimum of capital or other resources that may have been required as little as 50 years ago.

But leaving your job doesn't mean you automatically become a successful entrepreneur. If that were true, massive layoffs wouldn't lead to massive unemployment.

Successful entrepreneurs differ from other people — not in their abilities, but in their mind-set. They have internalized two fundamental commitments by making these two decisions:

- **Decision 1**

    To depend entirely on their own abilities for their financial security (because they realize that the only security is the security they create themselves).

- **Decision 2**

    To expect opportunity only by creating value for others (because they understand that this is the only unlimited source of economic opportunity).

The first decision is empowering because it frees individuals from the dependency of employment, in which they can only think of income in terms of having a job — a job provided by someone else. The second decision is empowering because it frees individuals from the dependency of entitlement, in which they think they are owed a living by others, regardless of whether they contribute anything first.

These two commitments — and the availability of new microchip-based tools — are the reasons why entrepreneurs in every industry are achieving extraordinary success while millions of other people around the world are losing their jobs.

If you have already made these two decisions, appreciate the significance of what you have done. Appreciate the psychological and creative advantages that your commitment to independence provides.

If you have not made them, there is no better time than the present. The nature of the microchip revolution at the present time is to eliminate jobs that are no longer productive, and to undermine the concept of entitlement which is the basis for the modern welfare state.

The sooner you make the two entrepreneurial decisions, the sooner you will increase your ability to navigate the unpredictable waters of global change, and chart your independent course to a more successful and satisfying lifestyle.

strategy 2

Get Away

From Bureaucracies

As Far As You Can

The bureaucratic culture is deeply ingrained in us. Throughout history, bureaucracies have characterized religious bodies, military forces, and all levels of governments. Business was the last to adopt it, spurred on by the Industrial Revolution, which brought with it huge head offices, factories, and assembly lines that required the efficient organization of masses of people. In fact, for the past hundred years, daily life in North America, Europe, and Japan was based almost entirely on the operation and expansion of huge industrial bureaucracies.

These organizations were built like pyramids, with many levels of managers and workers doing repetitive tasks. Until recently, this was the only structure that allowed for the orderly flow of communication and information up to and down from (but mostly down from) the people in charge.

Bureaucracies relied on conformity and loyalty.

In return, they gave you a form of security. In most major corporations, you went to work, and if you were relatively competent, you could be reasonably assured of a paycheck until you retired, and then a pension until death. In the industrialized countries, you grew up playing by the rules of a bureaucracy. And they were the only set of rules you knew. Then along came the microchip. Tasks that used to take entire departments to accomplish now take a computer keystroke. Information passes through phone wires, satellites, and computer cables at the speed of light instead of going from one in-tray to the next.

The collapse of large bureaucracies

Loss of Competence
Bureaucratic decision-making cannot keep pace with external changes.

Loss of Authority
Bureaucratic leaders appear confused, reactive, and self-serving.

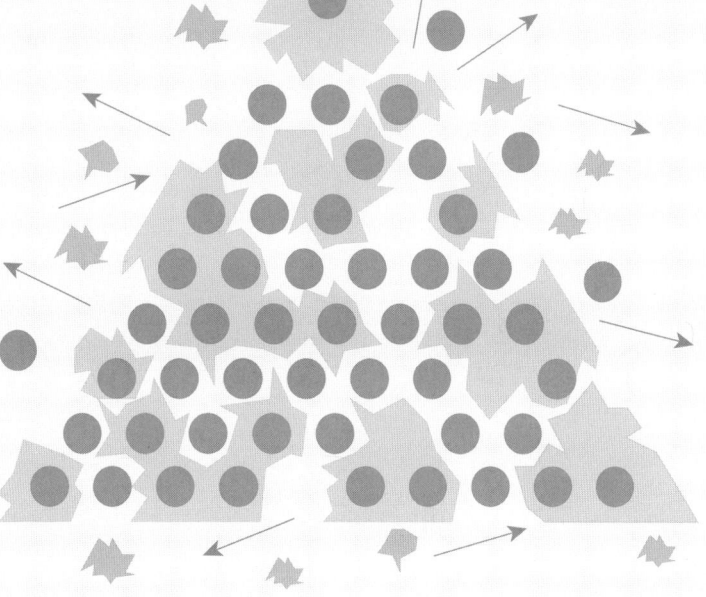

Loss of Security
Bureaucratic managers and workers are no longer sure about future employment.

Loss of Relevance
Bureaucratic organizations no longer serve public needs.

It's getting more difficult to make a living by shuffling paper.

Before microtechnology, bureaucracies allowed for the orderly flow of information to and from key decision-makers. Microchip-based networks now make this information flow possible in the blink of an eye. A single computer, equipped with a CD-ROM, modem, lan, or wan can now provide the same information as 10,000 bureaucratic workers. Ten years from now, the best computer systems may outperform as many as 100,000 bureaucratic workers. Computers are wiping out layers upon layers of management and workers. The pyramids are breaking up. The largest bureaucracies in the world, the Communist governments of Eastern Europe, have already collapsed under their own weight.

In business, management gurus call it right-sizing, re-engineering, streamlining, flattening the organizational chart, or paradigm shifting. But whatever they call it, the end result is that productivity has never been higher. Ditto for layoffs.

It's interesting how politicians are still trying to get elected on the promise of jobs, jobs, jobs. It's not going to happen. Over the next 30 years, most of the jobs and dependencies based on huge bureaucracies will be eliminated.

For people who have never learned any way of life other than life in a bureaucracy, it will be a slow and painful reckoning with a new reality. For people who learn how to create their own economic opportunities — by making the two entrepreneurial decisions and living by them — it will be the most exciting and successful period of their lives.

s t r a t e g y   3

Obey The

Laws Of The

Microchip

*M*icrotechnology affects every aspect of our lives: our financial prospects, our leisure time, our diets, our health — even our sense of emotional security. There have been other significant inventions throughout history — gunpowder, electric power, and the internal combustion engine, to name a few — which have fundamentally and universally altered the economic, political, and social landscapes, but no other technology continually reinvents itself like microtechnology.

The microchip is always getting faster and more powerful, allowing it not only to do the same tasks better, but allowing it to be applied to an ever-broader range of human activities.

This non-stop acceleration of computing power is precisely what is driving the fundamental shifts we are experiencing in cultural, political, financial, and interpersonal relationships on a global, local, and individual scale.

The sudden emergence of thousands of new enterprises and new types of occupations; the collapse of Communism and the disintegration of major bureaucracies in capitalist societies; the closing of thousands of industrial plants and factories; the deterioration of family life; the weakening of traditional values and customs; the general increase of anxiety, stress, and violent behavior around the world: All of these can be traced back to the increasingly rapid proliferation and speed of tiny devices called silicon microchips.

The Ten Laws of the Microchip

LAW 1
The speed of microchips doubles every 18 months.

LAW 2
The microchip favors individuals over organizations.

LAW 3
Cheaper information drives out expensive information.

LAW 4
Innovation is cheaper than competition.

LAW 5
The microchip disconnects money from time and effort.

LAW 6
The microchip disintegrates all human work except leadership, relationship, and creativity.

LAW 7
Each generation of microchips creates new organizational problems which can only be solved by the next generation.

LAW 8
Utilizing the microchip both encourages and requires endless automation and decentralization.

LAW 9
Economic growth and success result from increased participation in the global microchip system.

LAW 10
The microchip creates a global economy of constantly increasing returns.

That's because microtechnology is much more than just an explosion of useful machines and devices. It has created an entirely new way of approaching human work and life on our planet.

Although we are just in the early stages of the microchip revolution, already it is possible to discern ten "laws of the microchip" that are transforming the world around us:

**Law 1:
The speed of microchips doubles every 18 months.**

This is also known as "Moore's Law," after Gordon Moore, one of the inventors of the first microchip. In 1965, he predicted that the number of transistors (which provide the computing speed and power) would double approximately every 18 months as the transistors become exponentially smaller. He also predicted that the cost of chips would come down by about 30% per year. For the past 30 years, this prediction has been essentially accurate. And while he believes that this may slow down to a doubling every two years, this still represents an incredible, continual growth in microchip development. Trillions of dollars and the efforts of the most creative scientific minds on the planet are going into this effort.

This first law tells us to base our personal progress on the availability of ever-faster chips, and continually improved software to take advantage of greater microchip power. In this way, our personal capabilities will continue to be multiplied.

Law 2:
The microchip favors individuals over organizations.

It is no accident that the explosion of microtechnology has been accompanied by an explosion in entrepreneurial activity. Individuals are able to adapt more quickly than institutions to the availability of new tools, methods, and systems. Prior to the introduction of microchips, the most powerful technologies in society typically required vast investments of money and the employment of masses of workers to make them profitable. But with microtechnology, investment costs have plummeted to a few thousand dollars, and it requires only the creativity and skill of a single person — or a network of similarly motivated individuals — to create profitable new enterprises and industries.

Law 3:
Cheaper information drives out expensive information.

Microtechnology is making it possible to collect, organize, store, transform, package, and deliver vast amounts of information at a fraction of the cost than was possible only ten years ago. This explains why so many middle managers — who are essentially expensive conduits for the orderly flow of information from one level of staff to another — are seeing their jobs disappear. As the whole world continues to integrate itself with microtechnology at an exponential rate, massive amounts of information will become increasingly cheaper for everyone.

The third law tells us that the value of simply passing on this information will continue to diminish, while the value of manipulating information will increase. This includes filtering (separating the relevant from the unnecessary for specific purposes), interpreting, and transforming information into new and usable knowledge. This is resulting in millions of new entrepreneurial niches.

Law 4:
Innovation is cheaper than competition.

Competition used to be about scarcity and limitation — scarce materials, scarce technology, limited choice, and limited consumer markets. As a result, it was more economical to do almost anything to maintain competitive position than it was to improve the actual product. Microtechnology changes all that. It has become so easy to improve products and services that it is now the primary way to maintain a competitive edge. In fact, the most successful entrepreneurs on the planet are those who never compete in the traditional sense. As soon as the competition tries to imitate their existing success, they jump to a new level of creativity where there are extraordinary profits and no competition. This law underlies the need, and the reward, of continuous innovation.

Law 5:
The microchip disconnects money from time and effort.

The basic measuring stick for personal value within the bureaucratic world has always been the "work hour." Within this value system, those who worked long hours were considered more valuable and praiseworthy than those who didn't — irrespective of the results that were produced. This work-hour attitude measurement of value, however, is quickly coming to an end as the impact of microtechnology on all economic sectors radically changes our understanding of productivity. In the microchip-based economy of the 21st Century, work is becoming increasingly "results-based" — geared to opportunities, projects, and special situations where extraordinary financial results are possible if one achieves the best results. The question of getting paid in the marketplace no longer relates to the amount of time and effort put in, but rather to what results are achieved.

Law 6:
The microchip disintegrates all human work except leadership, relationship, and creativity.

Are there any areas of the work world where machines will never replace humans? The answer is a resounding "yes!" The three uniquely human capabilities are the ability to provide leadership in a world of infinite complexity, the ability to relate to other human beings in an infinite number of ways, and the ability to create an infinite number of new things. During the 21st Century, these are the only three areas of meaningful work that are safe from the encroachment of microchip-based tools, systems, and networks.

The application of microtechnology enhances the ability to create, relate to, and effectively lead others for those who are motivated, skilled, and focused within these three fundamental areas of activity. But at the same time, microtechnology will continually disintegrate and eliminate forms of work that are not based on valued leadership, increased creativity and expanded relationships.

Law 7:

Each generation of microchips creates new organizational problems which can only be solved by the next generation.

Large bureaucratic organizations in any industry are continually forced to introduce new technologies to increase their productivity. What invariably occurs, however, is an increase of complexity within the organization. The application of new technologies, while addressing existing problems, creates others which require higher levels of technology, technical expertise, and spending to solve. The largest organizations, therefore, must be in a perpetual state of increasingly costly reorganization.

Law 8:

Utilizing the microchip both encourages and requires endless automation and decentralization.

A fundamental principle of microtechnology is "distributed intelligence" — meaning that the greatest organizational progress is achieved when the greatest number of people are involved in decision-making and creativity. Power and responsibility must continually be taken away from the organizational center and be dispersed as widely as possible through automated systems of communication. In most industries, entrepreneurial organizations are able to do this much better than are large bureaucracies.

Law 9:
Economic growth and success result from increased participation in the global microchip system.

In order to grow, many companies are finding today they must become more based on microtechnology. The goal is to combine the greatest amount of innovation with the greatest amount of automation. Smaller, enterpreneurial organizations are able to increase their useful participation in the global system of microchip technology more quickly than are larger organizations. Entrepreneurs can make changes in weeks and months that take bureaucrats years to achieve.

Law 10:
The microchip creates a global economy of constantly increasing returns.

As microtechnology has freed up the creative capabilities of large numbers of individuals and whole organizations, it has become possible to achieve return on investments and growth of assets exceeding 30% or more per year. It is now possible to operate within an economy of ever-increasing returns. This is more viable, however, for smaller, entrepreneurial organizations than for larger organizations, which have higher costs, are slower to utilize technology, and are more resistant to entrepreneurial innovation.

strategy 4

Let The New

Technology

Come To You

*d*on't panic!

After realizing how thoroughly immersed we are in a global electronic economy, many people today are worried that they are falling behind in the area of technology, and that others — competitors and enemies — are gaining a superior advantage that they will never overcome.

This is called Technological Panic, and it's one of the most effective ways that the makers of new technologies sell their products — by scaring us with the terrible thought that, because we don't have their product, the world is passing us by. But this is just an old sales ploy that's probably been used since people lived in caves.

On one level, all of this microchip stuff is like attending a three-ring circus. There are more clowns than tigers, the action never stops, and there's never a dull moment. There's a lot of deception and trickery, and there are many outrageous claims. We've got to constantly be on the lookout for con artists and pickpockets. Whenever we start to get too anxious about technological progress, we should remind ourselves to laugh, relax, and just enjoy the show — and at the same time, keep a tight grip on our priorities and our wallets.

As long as we maintain our personal confidence and clarity about our own specific and unique future, the world of the microchip isn't going anywhere without us.

The strategic approach to technology

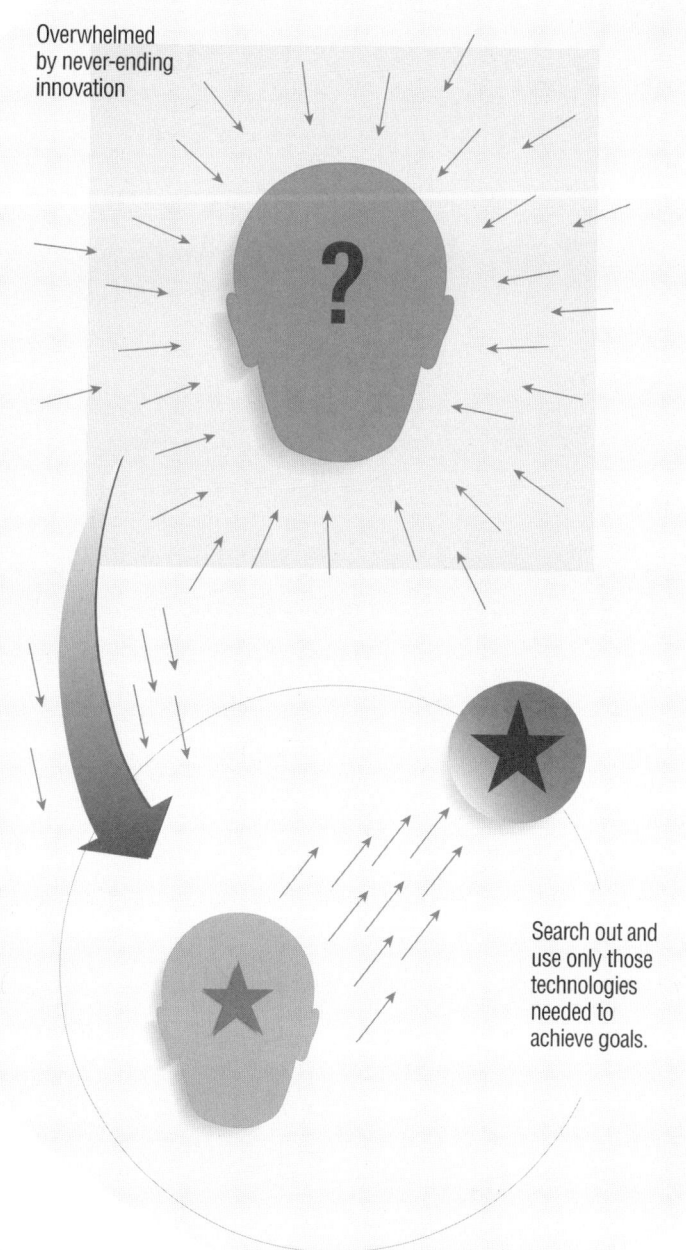

Just because there's something "new and improved," it doesn't mean you have to have it. If something won't help you work smarter, then there's nothing smart about working with it. With that in mind, here are several ground rules for achieving peace of mind in a world of never-ending technological breakthroughs:

- Generally, new technology isn't useful for anybody until it's useful for everybody; that is, when everyone can use it easily, economically, and without special training.

- The people who create new technologies are usually just guessing, gambling, and hoping; most of them don't really know any more about the future than you do.

- The vast majority of new technologies fail in the marketplace; they're mostly experiments on the way to something better.

- The third generation of a technology is usually vastly superior to the first. Wait for the grandchildren of any new device or system.

- Whether a new technology is useful or not depends upon the goals and purposes of the individuals using it.

- The most outrageous claims for new technologies are often made by technology makers that are in financial trouble. They're hoping that consumers will bail them out.

The biggest adjustment that we need to make in relationship to technology is simply to accept that we now live in a world that is based on technology — where a never-ending flow of new technologies is a normal part of everyday life. Countless devices are created that we never know about. And much of what is available today will be obsolete two years from now.

No one can keep up with all of the changes. No one can possibly stay ahead of this game. No one can control the future direction of even a small part of where technology is going.

So take it easy, take your time, and keep your cool. All of the technology that you will ever need for the rest of your life will come to you when you truly need it.

strategy 5

Give Children An Entrepreneurial Attitude

*i*t's a lot tougher being a kid today than it was 25 years ago — mainly because most older people can't provide them with good advice and direction about the future. If children are confused today, it's because many of the adults who are supposed to be role models and teachers are even more confused than they are. It used to be that a good education was your best bet if you wanted to have a secure, stable future with a good job. Not any more.

The public education system appears to be stuck in a structure that continues to prepare our children for jobs that are disappearing as quickly as are the bureaucracies that used to create them.

Parents, teachers, and students are worried. And rightfully so. The transition to a global economy fueled by microtechnology is fundamentally changing the rules of the game, and the public education system may be one of the last sectors in society to catch on. Few of today's teachers, principals, and politicians responsible for education have any idea of the requirements of the entrepreneurial companies that will increasingly be the source of new jobs. And the ones that do, have little ability to change the system. Being a good student today gives no indication of a child's economic future. Thousands of university graduates are unemployed or taking on menial jobs, and even the teachers and principals who lose their jobs have few options outside the public education system. This often leads to a bleak view of the world that is being passed on to our children through the schools. But in reality, the future is far from bleak.

The crisis of the educational system

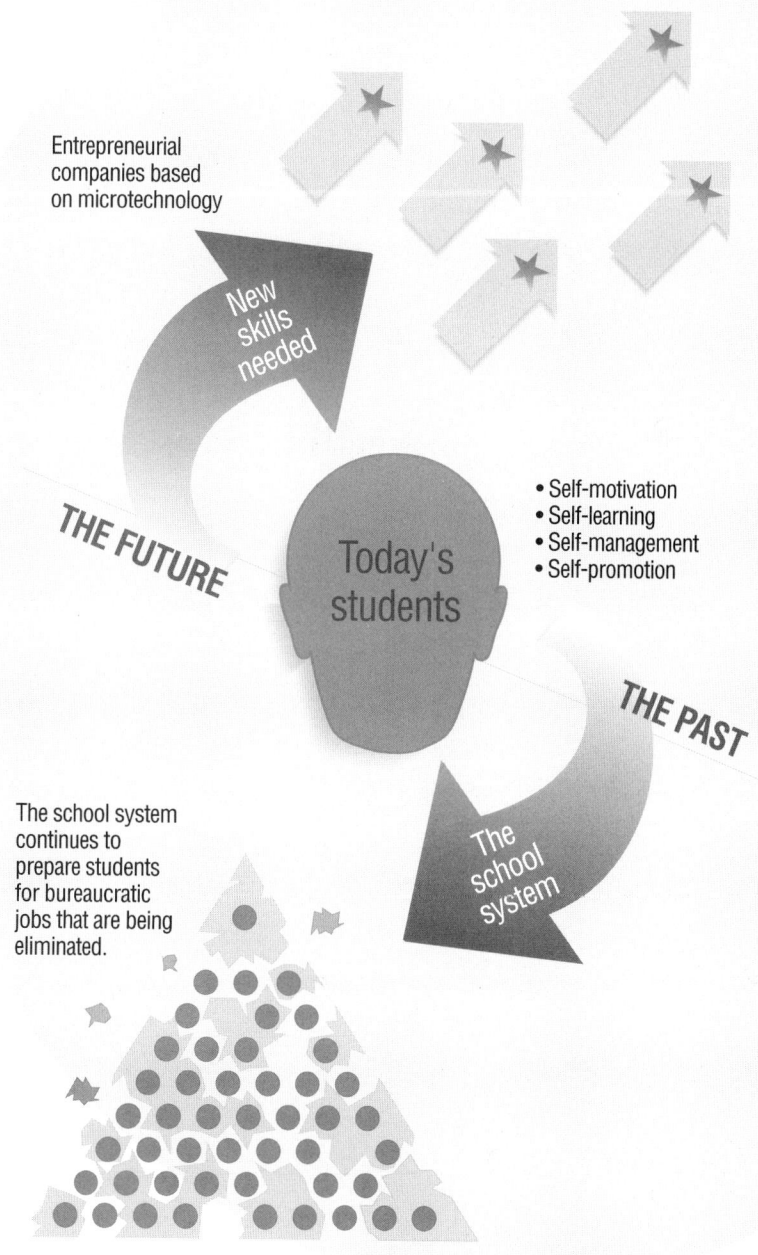

Entrepreneurial companies based on microtechnology

New skills needed

THE FUTURE

Today's students

- Self-motivation
- Self-learning
- Self-management
- Self-promotion

THE PAST

The school system

The school system continues to prepare students for bureaucratic jobs that are being eliminated.

Virtually unlimited economic opportunities are unfolding for those individuals who understand the power of microtechnology, are willing to adapt on a continual basis, and are able to contribute as members of cooperative teams.

Small, entrepreneurial companies require that employees themselves have an entrepreneurial attitude, with a willingness to be paid on performance rather than education level, social connections, or seniority.

It's not only *what* our schools are teaching children that needs to change, it's *how*.

The "three Rs" will continue to play an important role in providing the groundwork for all future learning at any age. The problem is that in addition to this knowledge, the current education system promotes an attitude and a set of habits based on obedience, dependence, and loyalty — when it should be encouraging self-reliance, creative thought, and initiative.

These require four specific skills:
- **Self-motivation**
  Setting and achieving goals in a systematic manner.

- **Self-learning**
  Acquiring new knowledge and skills without the assistance of a teacher.

- **Self-management**

  The effective organization of time and money.

- **Self-promotion**

  The ability to present oneself to others in a way that creates new opportunities.

The biggest disservice we can do to our children is to wait for the public education system to change in time to prepare them for the economic challenges and opportunities that lie ahead. The best thing we can do for them is to take the lead and become role models — to encourage our children to become entrepreneurial by adopting these attitudes, habits, and skills ourselves.

strategy 6

Decide To Be

Happy Right Now

And Live With It

Thomas Jefferson caused a lot of problems when he said that we all have an inalienable right to "life, liberty, and the pursuit of happiness." Life, certainly. Liberty, no disagreement. But it's that last one, the pursuit of happiness, that causes so much unhappiness. The problems lie in the notion that happiness is something that needs to be pursued. But happiness is not a pursuit. It's not something out there, nor is it something external to us. Happiness starts with the most personal commitment that any human being can make.

First, we decide to be happy.

Individual happiness begins the moment that any person, regardless of his or her circumstances, decides to be happy. A person can be in miserable conditions, lacking advantages and opportunities, and still decide to be happy. The moment that a person decides to be happy, he or she immediately begins to gravitate to other people who have made the same decision. The decision to be happy communicates itself to the world, and all of the other things in the world that are already happy — people, places, activities, and circumstances — begin to communicate back. As we move forward into the 21st Century, and as the promise and performance of microtechnology become greater, we are going to be told constantly that the next technological breakthrough will be the one that finally makes us happy. Don't believe a word of it. None of these new methods and machines are producing any more individual happiness than anything we have created in the past. The decision to be happy for the rest of our life is a decision that each of us can make right now.

Happiness creates its own environments

RELATIONSHIPS

CIRCUMSTANCES

The decision to be happy

CAPABILITIES

OPPORTUNITIES

s t r a t e g y   7

Choose Creativity

And Stop

Complaining

a fundamental part of intellectual and emotional maturity as a human being is the recognition that the world we live in was not designed with any of us in mind.

The world is there for each of us to utilize, but it's not there to satisfy any one of us.

This is especially true as we move deeper into an age governed by microtechnology, where there is a constant disruption of the existing order, where old structures of security are disappearing, and where all of us are continually confronted with changes we didn't ask for. And so we are left with a fundamental choice about the direction of our lives.

We can complain about the world. Or we can choose creativity and transform our circumstances.

It's unfortunate that so many grown men and women reject the path of creativity, as evidenced by the innumerable complaints that fill each day's news reports. These people spend many of their waking hours over an entire lifetime complaining about unfairness, deprivation, inequality, and a hundred other real or imagined injustices. Yet all complaining does is to make them even more miserable.

Complaining only reinforces the strength of the perceived problem. It's a non-creative activity that prevents people from seeing any alternatives to their present situation.

Transforming obstacles into strategies

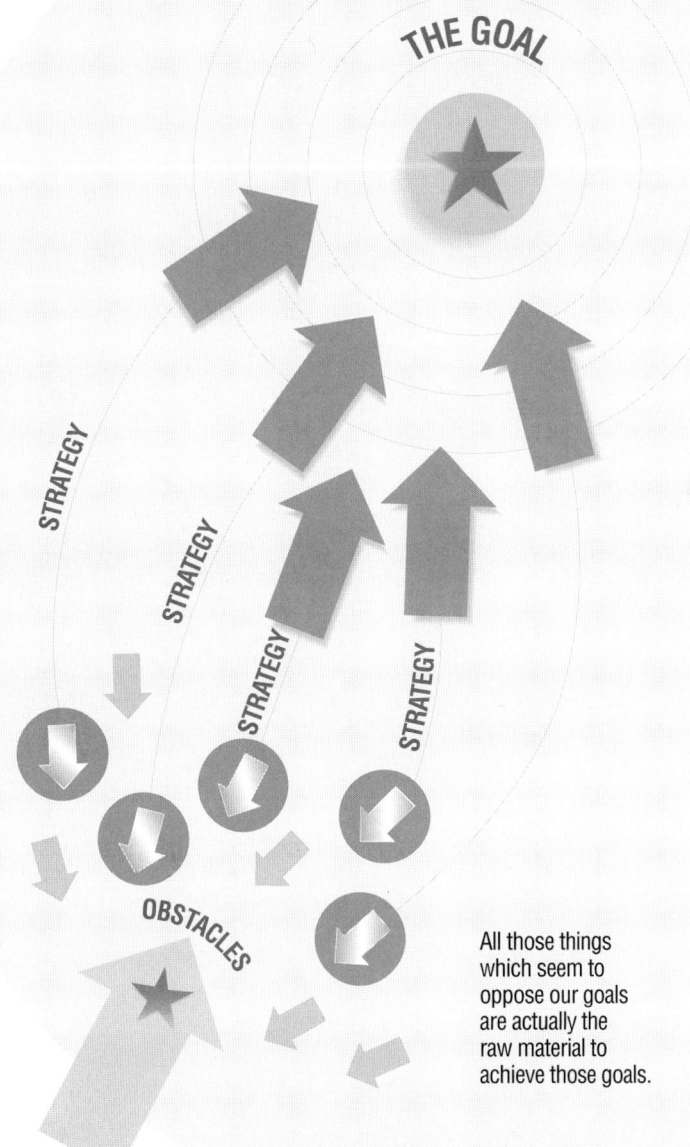

All those things which seem to oppose our goals are actually the raw material to achieve those goals.

When people complain, they give up some of their personal power to outside forces. Eventually, they give up all of their power and become totally powerless, entirely through their own choice. The act of complaining also attracts other individuals who are complainers, so that in a short time the entire human environment is negative, pessimistic, cynical, and resistant to improvement.

This is what is known as a "culture of complaint," in which individuals — sometimes highly educated, privileged, and talented — squander their entire lifetimes competing to see who has the worst grievance against the world. Complaining makes it impossible to escape from the condition of chronic personal dissatisfaction.

Creators don't complain; complainers don't create.

Creativity works in a totally opposite direction to complaining — a positive direction. Creativity, as a lifetime habit of thought and action, leads to a permanent escape from negativity, regardless of one's circumstances. Everything we need to be creative is around us all the time. All we need is a clear vision of what we want to create — a powerful goal — and then a clear identification of everything that stands in the way. The creativity lies in transforming the obstacles into specific decisions, communication, and actions that lead directly to the vision.

All those things which seem to oppose your goals are actually the raw materials for achieving them.

Through the act of creating, there is excitement, hope, and the emergence of new opportunities for learning and growth. The person who is a creator can change, alter, transform, and improve his or her situation in life — regardless of what problems may be afflicting the general population. Unlike complaining, creativity always increases personal power. The more that a person creates, the more personally powerful he or she becomes throughout life.

Just as complainers attract complainers, creators also attract creators. This is why certain groups in society are highly productive and successful, while others are stagnant and self-destructive.

Choosing to be creative is similar to the choice to be happy, in that all your outward attitudes stem from an internal decision. But whereas the decision to be happy is directed to the present, the decision to be creative is directed to the future.

Choosing creativity leads you onto a path of constant growth, higher consciousness, inspiring relationships, and enormous satisfaction.

All future situations and relationships become increasingly positive and rewarding. Creativity makes no guarantees to anyone. What it does offer are constant surprises, challenges, new opportunities, and breakthroughs.

Complain or create. It's the choice of a lifetime.

s t r a t e g y   8

Focus On

Progress And Forget

About Perfection

*P*erfection as an objective was instilled in our culture by organizational institutions such as religions, governments and the military. It was part of the way these bureaucracies acted as stabilizing forces in society and worked to slow down the rate of change. That was how they endured even as people changed. For thousands of years, the pursuit of perfection actually helped to simplify our lives. Despite the fact that we knew we could never achieve it, there was some measure of comfort in knowing that some ideal existed.

Today, constant innovation is throwing all that out the window. Anyone looking for an ideal to emulate is faced with a target that is moving so quickly, it can't be defined at all. Individuals and organizations which continue to hold themselves accountable to any standard of perfection are the least able to cope with change, because they continually focus on deficiency and failure. Those who focus on progress, on the other hand, are the best able to take advantage of opportunities as they arise because they have gained the confidence that comes with acknowledging their accomplishments.

Focusing on perfection makes people self-conscious and judgmental of others, and it discourages experimentation and risk-taking. Focusing on progress encourages individuals and organizations to build interpersonal relationships that are open, trusting, and supportive.

People in perfection-based corporate cultures are always on the defensive, and they seem to be most interested in rationalizing

Choosing achievement over efficiency

why initiatives won't work. People in progress-based corporate cultures are more interested in assisting others to achieve extraordinary goals, which also creates the greatest opportunities for themselves.

As the world continues to change at an accelerated pace, it is becoming increasingly important for both individuals and organizations to focus on continual progress, and forget about ever achieving perfection.*

* For further reading, see *Learning How To Avoid "The Gap."* Dan Sullivan (1999). The Strategic Coach Inc.

Where is your focus?

Focused on Progress	Focused on Perfection
Set specific, realistic goals.	Goal is always perfection.
Celebrate achievements and build on them.	No time to celebrate because you are "not there yet."
Lead by instilling confidence.	Lead by instilling fear.
Measure progress by how far you have come, not by how far you have yet to go.	Afraid that if you don't hold yourself to the standard of perfection, you'll stop striving.
Use perfection as a motivator, not a measuring stick.	Hard on others, but hardest on yourself.
Recognize that perfection is a moving target.	Performance is never "good enough."
Experience a sense of confidence, success, capability, accomplishment (successful and happy).	Unable to recognize achievements or enjoy success to the fullest (successful, but unhappy).

strategy 9

Focus On Habits And Forget About Discipline

"If only I were more disciplined." How many times has this thought gone through your mind? If you're like most people, the chronic self-criticism of not having enough discipline is a recurring source of discomfort and regret throughout life. The concept of "discipline" is a major cause of personal guilt. It makes people feel continually negative about themselves because they never feel they have enough discipline, and it prevents them from deriving satisfaction from the real progress and growth that is continually taking place.

The time has come to get rid of the concept of discipline.

Discipline is a lot like perfection. It is based on the belief that there is a fixed way of doing things that other (superior) individuals have already mastered, and that the main task in life is to bring one's behavior into conformity with the standards of perfection established by higher authorities. As long as our society was primarily based on bureaucracies, the pursuit for greater self-discipline made a lot of sense for those who were trying to secure a permanent place within these organizations. Within the rigid hierarchies of these institutions, the pursuit for greater self-discipline was encouraged, honored, and rewarded. Growing up, our parents told us that we needed self-discipline to succeed. In school, our teachers told us we needed self-discipline to get good grades. At work, our bosses told us we needed self-discipline to get the next promotion.

All that may have been good advice in the old world of bureaucracy, but not in the new world of microtechnology.

Everyone is completely disciplined

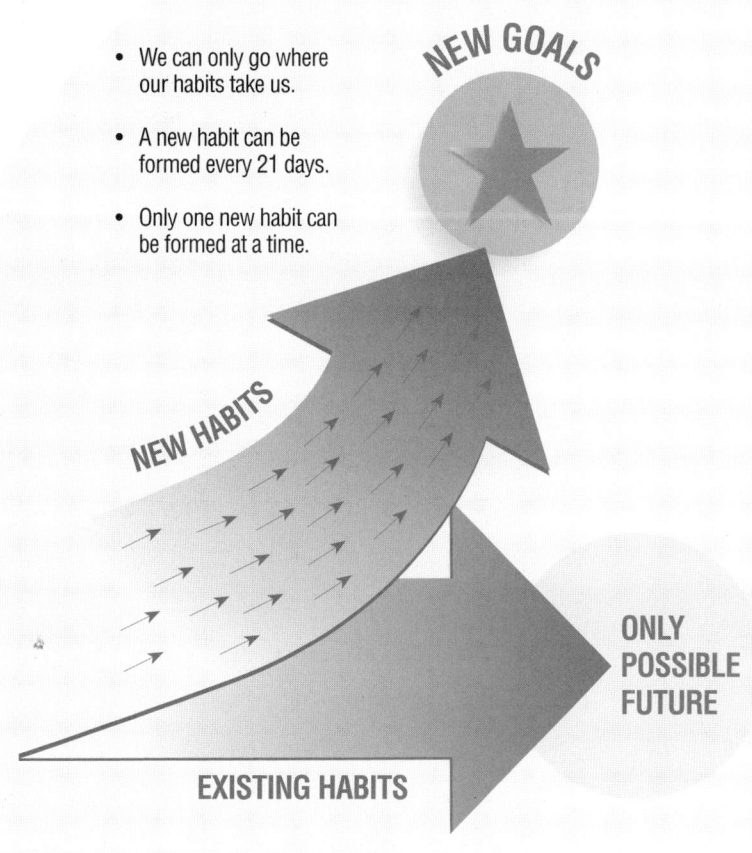

- We can only go where our habits take us.
- A new habit can be formed every 21 days.
- Only one new habit can be formed at a time.

NEW GOALS

NEW HABITS

ONLY POSSIBLE FUTURE

EXISTING HABITS

What are the habits that make up the discipline?

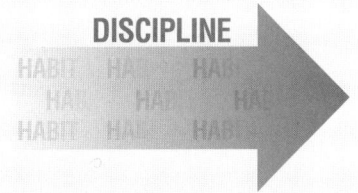

DISCIPLINE

ALL HABITS = DISCIPLINE

The microchip has no time for self-discipline. There's too much opportunity for personal growth all around us now to waste our time, energy, and attention being concerned about our deficiencies. There's no longer any time to waste feeling guilty about not living up to the obsolete standards of some bureaucratic authority.

The truth is that all of us are already 100% self-disciplined — to our existing set of habits.

Instead of focusing on self-discipline, we need to take a closer look at our habits. We have work habits, relationship habits, health habits, leisure habits, and so on. In fact, each of us has a full set of habits that regulate all of our behavior. The only thing making one habit good and another one bad is if it supports or defeats our goals.

Successful people have successful habits; unsuccessful people have unsuccessful habits.

Do your habits work for you? Or against you? Those are the only value judgements that make any sense when looking at your habits — or anybody else's for that matter. Personal progress, then, is the continual process of aligning your habits with your goals at any given time and in any given environment. If a habit doesn't work for you, you can change it. It has nothing to do with self-discipline.

The most valuable habit you can acquire in a world of constant technological change is the habit of changing your habits.

The whole world around us is changing faster today than during any other period in history. Therefore, your circumstances will change. Your goals will change. And your resources will change. So as you resolve to make changes in your personal and professional life, here are four points to keep in mind:

- Your brain doesn't care which habits it reinforces. Good habits feel just as natural and comfortable as bad habits.

- It takes 21 days of daily repetition before a behavior becomes a habit.

- You can expect to consciously change a maximum of one habit at a time — yet changing just one habit automatically changes several other habits at the same time.

- Over a three-year period — at 21-day intervals — you can change 52 fundamental habits, one at a time, in all areas of your life. This would dramatically improve the results you could achieve.

Focusing on habits rather than discipline relieves the burden of self-guilt and provides a constant increase in self-confidence. You are the expert on all of your own habits. You know which ones work for you and which ones must be changed and improved. You can begin changing your first habit today.

strategy 10

Focus On Your

Strengths And

Delegate Your

Weaknesses

*h*ow many times have you been told to work on improving your weaknesses? That may be the conventional wisdom which many of us learned as children, but it is also bad advice. It is responsible for more wasted time, talents, opportunities, and lives than almost anything else.

Fortunately, successful entrepreneurs, as well as top scientists, artists, athletes, and entertainers throughout history have proven that you shouldn't take conventional wisdom too seriously. These people have achieved greatness by focusing on their areas of strength. Take Frank Sinatra, for example:

Frank Sinatra did not move pianos. Frank Sinatra sang.

In other words, he focused on his strengths, and he surrounded himself with a support team of individuals with complementary skills and expertise who contributed to the overall effort needed to create a superb performance. This applies in every other area of human endeavor.

Everybody has a natural aptitude in certain areas. No matter how hard you try, it is highly unlikely that you will ever be more than average in areas where you simply do not have an aptitude. As a result, continually working on your weaknesses undermines your self-esteem since you will focus mostly on your deficiencies. If you spend your time working on your weaknesses, by the end of your life you will have a lot of very strong weaknesses — and you will feel just as deficient about yourself when you are old as you did when you were young.

Concentrate only on your strengths

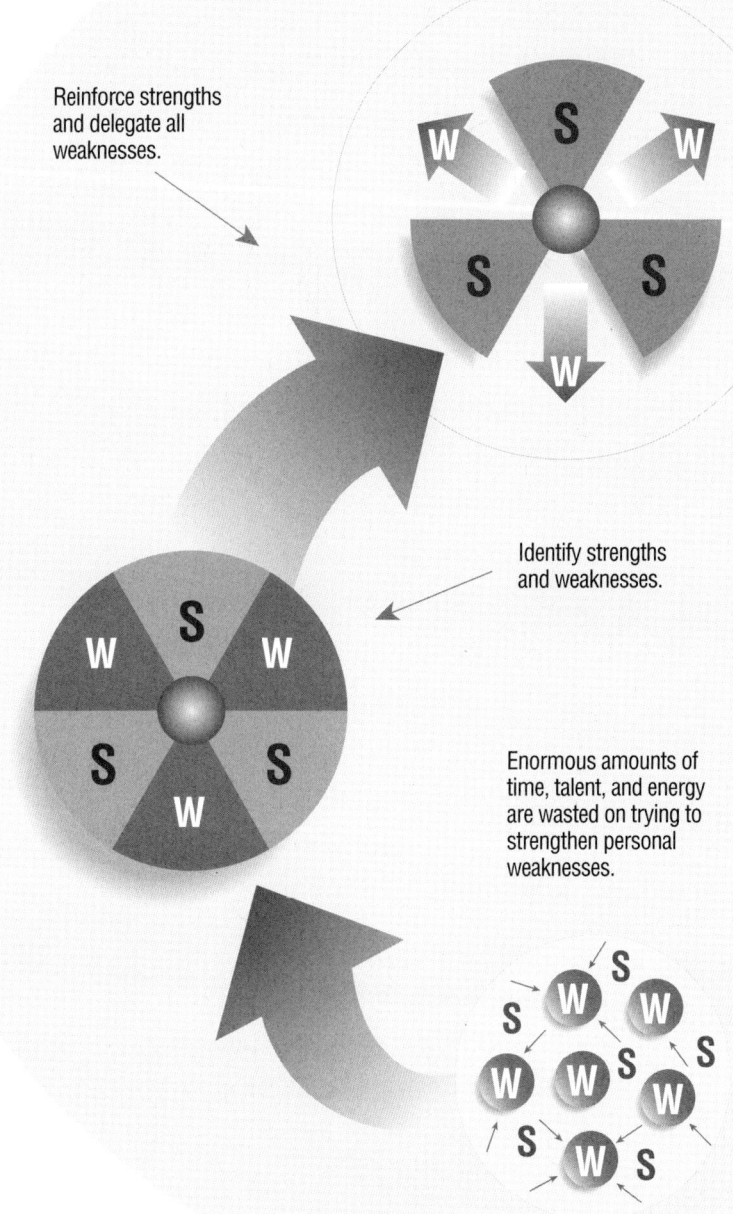

Reinforce strengths and delegate all weaknesses.

Identify strengths and weaknesses.

Enormous amounts of time, talent, and energy are wasted on trying to strengthen personal weaknesses.

If you try to become good at everything, you will never experience the immense satisfaction that comes with being superb at any one thing. On the other hand, developing your natural talents is self-rewarding and motivating, and you will continually realize higher and higher levels of ability, achievement, and success. Not only is this more important than ever, it is also more feasible. With today's vast electronic networks, we now have the ability to link up with the best people in the world.

Increasingly, superb results come not from a single individual, but rather from teams of individuals, all working within their own distinct areas of expertise toward a shared goal or vision.

If you're already an entrepreneur, you have leveraged one or two strengths into a business. The better you become at those, the more successful your business will be. So what should you do about all the other skills you need to run your business more efficiently and effectively?

Delegate.

Stop trying to strengthen your weaknesses. Find people who are naturally talented in all those areas which you will never be good at anyway, and concentrate on strengthening your strengths.

Do you want your life to be about mastery or mediocrity?

Mastery	Mediocrity
Delegates tasks and responsibilities to others with a natural aptitude in those areas	A "rugged individual" who insists on doing everything him- or herself
Focuses on developing strengths to genius level	Tries to strengthen weaknesses
Builds a lifetime of activity around superior skills that also generate excitement and energy	Not enough time or energy to spend developing strengths to genius level; too much else to be done
Ability to clearly see paths for growth and to take abilities to the next level	Hits a "ceiling" where life becomes too complex to take advantage of new opportunities
Never-ending earning and growth potential	Earning potential and growth are limited by individual's time and energy

s t r a t e g y   11

Improve Your

Mastery Of The

English Language

everybody's now in the business of selling. This should come as no surprise to entrepreneurs, who have long learned the importance of selling a benefit to customers, a business plan to bankers, and a vision to employees. But it's not just entrepreneurs who sell. Someone looking for a job has to sell his or her experience and capabilities to a prospective employer, politicians must sell a platform to the electorate, and teachers have to get their students to "buy into" a reason to learn.

Everybody "sells" something, and the most important tool we have to sell with is language. Language skills are the foundation of communication skills, which are the foundation of selling skills.

Since the beginning of time, the most powerful, influential, and successful people have always been those with the greatest command of the written and spoken language.

The only thing different today is that it is more important than ever to have superior language skills. Why? Because in today's Information Age, value is based on your ability to receive new information and knowledge and to create new information and knowledge. This is dependent on your ability to use language.

Improve your mastery of the English language.

What about the argument that because we live in a global economy, English-speaking people should learn a foreign language?

Why English is now the global language

AMERICAN CULTURE
DOMINATES
GLOBAL MARKETS:

- Film
- Television
- Music
- Fashion
- Sport
- Lifestyle

The Microchip

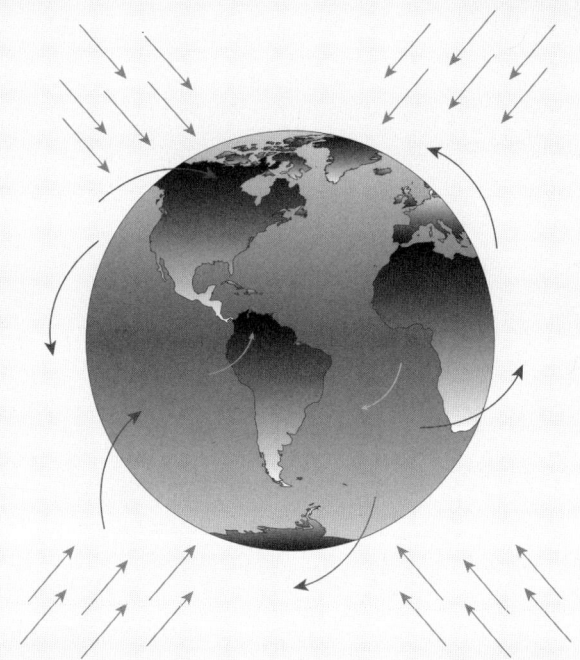

WORKING
LANGUAGE FOR:

- Science
- Technology
- Business
- Politics
- Economics
- Journalism
- Entertainment
- Academics

LANGUAGE OF
INFLUENTIAL
DEMOCRACIES:

- United States
- Great Britain
- Canada
- Ireland
- Australia
- New Zealand

Depending on the viewpoint of the person who is making this argument, different languages are suggested as the best ones to learn: Japanese if one argues that Japan is taking over corporate America; Chinese if one attaches great importance to the fact that China has the biggest market; or Spanish if one expects Latin America to be the next dominant force. But the truth is that you have more to gain by improving your mastery of English than by learning any other language. And this is true whether you live in an English-speaking country or not.

Why English?

The continual spread of microchip technologies is creating a global village that transcends national borders and unifies societies. As the cost, speed, and ease of communication across oceans become more like communication across the street, language itself becomes the last barrier to overcome. Unless a perfect, simultaneous translator becomes a reality, a single, primary language is inevitable. It is increasingly clear that English is it.

Since the end of the Second World War, English has become the dominant language of science, technology, and business.

As a result, it has also become the dominant language of politics, economics, journalism, entertainment, academics, and even the Internet. There are many theories as to why English is dominant, and probably all of them are at least partially correct:

- English is remarkably free of dialects compared with other major languages, making it easier to communicate with English-speaking people all over the world.

- English is relatively easy to speak, and even if one does not know the language perfectly, one can still communicate reasonably well (unlike languages such as French, which must be spoken nearly perfectly if the speaker hopes to earn any respect, or Asian languages which can take years of study to be able to communicate at all).

- For the past hundred years, America has been the source of the greatest number of scientific and technological breakthroughs — which have all been announced and disseminated in English first.

- The United States became the dominant world power just when mass communications emerged to help "export" American films, music, and television around the world. (Even the worldwide popularity of British music became possible because of its acceptance in America.)

- With more democracies in the world being English-speaking (United States, Great Britain, Canada, Australia, and New Zealand), English has become inexorably linked with the concepts and structures of democracy — a powerful and highly desirable ideal for everybody on the planet.

While all of these are historic reasons for the current dominance of the English language, there is an additional factor that will virtually ensure English will continue to dominate for some time: the global emphasis on the development of software.

The best computer software programs are written primarily for the American market, and the greatest breakthroughs in software development are taking place in the United States. In order to keep up with and benefit from the productive capabilities of the latest computer technology, it is essential to speak English.

This does not mean that learning a second language isn't helpful. On the contrary, learning the language of the market you operate in can help you understand the culture of that market — always a key factor when doing business in a foreign country.

Learning a language that is related to English can actually help you improve your mastery of English, while learning any language can broaden your mind by teaching you other ways of thinking.

But the end goal in any case is improving your mastery of the language of the world — English. Reading, writing, listening, speaking, selling: Your success in the global economy, of which we are all a part, is based on your ability to understand and use English effectively.

"One out of four of the world's population speaks English to some level of competence. Demand from the other three-quarters is increasing. English is the main language of books, newspapers, airports and air-traffic control, international business and academic conferences, science technology, diplomacy, sport, international competitions, pop music and advertising."

Source: The British Council web site, http://www.britcoun.org

strategy 12

Buy Peace Of Mind

In A Debtor's World

With 20% Savings

*h*ow many people you know have a mortgage, a car loan, possibly a second-mortgage for renovations, and a line of credit that never seems to get paid down? Of those people, how many are stressed out, or so caught up in making enough money to meet this month's payment that they can't even think about anything else. Or trapped in unfulfilling and uncertain jobs because they are terrified of losing their paychecks and benefits? That's what debt does to people.

Borrowing has become a permanent way of life for many people.

During the last 30 years, borrowing money for anything and everything became the norm. It was even considered smart. With inflation running high, it actually made sense to pay for things at a later date with devalued dollars.

This way of thinking was premised on the continued growth of the industrial economy, in which there was a constant creation of ever-higher-paying jobs. As long as the overall population was employed, there would be no problem in paying off the debts that were being incurred. Even recessions didn't frighten people away from debt. The economy, we all reasoned, would always bounce back.

It wasn't just individuals who got caught up in this borrowing epidemic. Governments at all levels ran up massive tabs that will have to be paid back by many future generations.

Freedom where others have little

THE HABIT OF 20% SAVINGS

- Able to make strategic decisions
- Able to take abundant free time
- Able to acquire new capabilities
- Able to be independent of any situation
- Able to concentrate on long-range goals

Government is a powerful teacher — for good or bad — and for the past 30 years the leaders of most governments on the planet have been teaching their citizens to be economically careless and foolish.

Nobody, it seems, counted on the fundamental restructuring in the economy which we are now seeing.

The fact is that since 1971, when the United States disconnected the dollar from the gold standard, the world has increasingly become an interest-rate-driven economy over which governments have little control.

Combined with the rapid growth in computer technology — which enables billions of transactions to be made instantaneously anywhere in the world — we now have a global "electronic" economy that is 50 times greater than the total amount of actual currency in circulation. The result is an economy that is extremely volatile.

The business cycles may still be identifiable (in hindsight), but they will likely occur faster and with sharper, more unpredictable swings. You can no longer count on interest rates staying low, or staying high. You can no longer count on continued growth of blue chip stocks. You can no longer count on the stability of many industries, nor the permanence of many businesses. And you can no longer count on a job. The only thing you can count on is your invested savings.

Savings buy you freedom:

- Savings give you the freedom to make strategic business and career moves during transition periods in the market, while competitors have to be more concerned with current bottom-line performance.

- Savings help you "buy" the considerable amount of free time that is necessary for mental, emotional, and physical rejuvenation, and for strengthening relationships with family and friends.

- Savings allow you to take time to learn new skills, investigate new opportunities, and acquire new technological capabilities.

- Savings mean that you are not dependent on any one job, boss, or client, and let you walk away from any situation or relationship that doesn't enhance your life or business.

- Savings allow you to focus entirely on the most important activities without worrying about short-term financial obligations.

Savings — not current revenue or borrowed money — are what make all these things possible. For this reason, put aside 20% of all income for long-term savings, and use savings as the primary yardstick for entrepreneurial success.

The habit of saving can be established immediately.

While your savings objective should be at least 20% of your income, it may not be possible to put aside this amount right away. In this case, begin with some set percentage, even if it is as low as 1%, and gradually increase your savings rate over time. This is best done with a game plan and the assistance of a qualified financial planner. Your savings can take many forms, such as certificates of deposit, real estate, mutual funds, stocks, bonds, or shares.

With respect to personal money management, the wisdom of previous generations is more important and appropriate now than at any other time since the Great Depression. They believed that the only way to buy something was with your own money. It may take longer to accumulate material things when you place a higher priority on saving rather than spending, but life is undeniably more pleasant and stress-free.

This lesson is an especially difficult one for entrepreneurs. They are used to reaching for higher and higher goals, which appear more easily attainable with capital — usually borrowed. Given today's economic reality, however, a solid base of savings is more essential than ever for continued success.

The price of freedom is 20% of your income in savings, every year.

strategy 13

Eat, Sleep,

Exercise, And

Meditate

*t*oo many people forget that they actually have a physical body until it is too late. During the past 20 years, there has been a dramatic increase in stress-related diseases, psychiatric disorders, and the consumption of both legal and illegal drugs. You don't have to think too hard to understand why.

The never-ending change which we are coping with during the microchip revolution is producing the highest level of stress the general population has ever experienced. Mass communications, including television, radio, the Internet, e-mail, advertising — even the telephone — require us to constantly process information and to maintain a very high level of mental activity whether we want to or not. Every one of us understands only too well the meanings of "information overload" and "information anxiety."

The need to continuously learn how to use new technologies makes it difficult for us to feel any sort of mastery or control over our surroundings. Even everyday tasks such as programming the VCR have defeated the majority of otherwise intelligent and successful people.

The technological world we live in can wear us out very quickly — unless we act strategically to strengthen our mental, emotional, and psychological well-being on a daily basis.

Every one of us must develop a personal strategy for maintaining our overall personal health. Eating well, sleeping well, daily

Physical health in stressful times

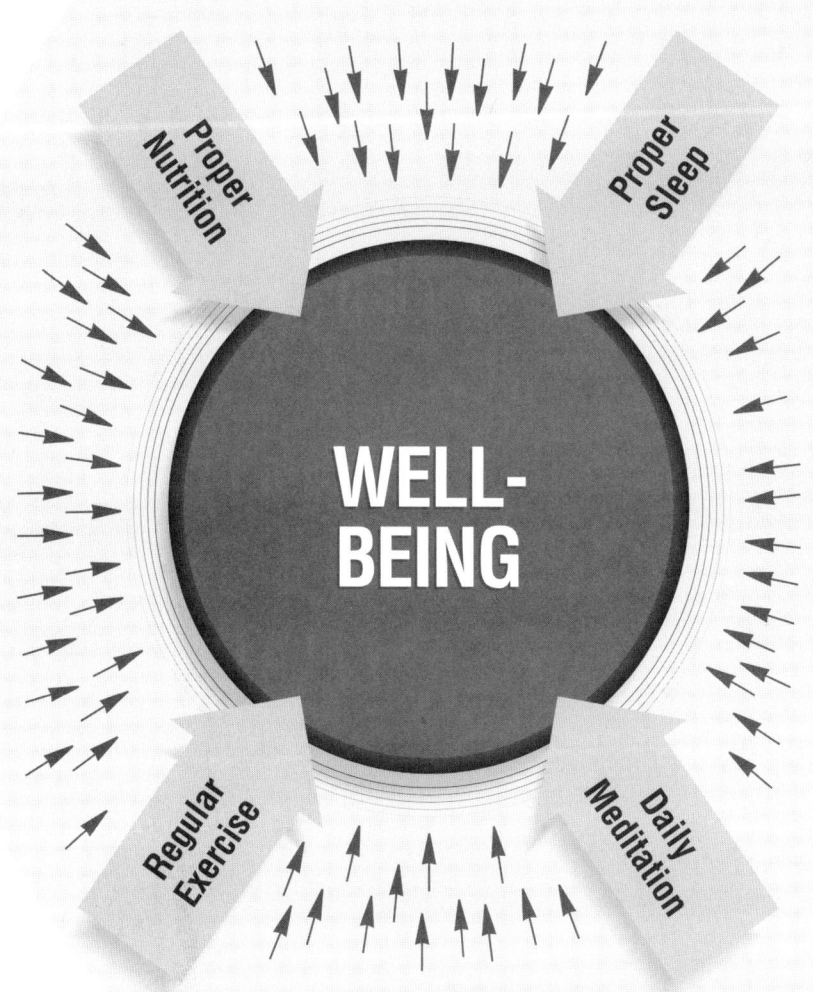

exercise, and meditation are the most important elements of this strategy. It's obvious that eating well and sleeping well help our bodies resist disease and maintain a sharpness of the mind. But then why do so many people continually punish their bodies with poor sleeping and eating habits?

The answer is chronic fatigue. They've allowed themselves to become so worn down that they've forgotten what good health really feels like. To regain that feeling, we must exercise regularly, starting today.

The benefits of general, moderate daily exercise are universally acknowledged. Exercise strengthens the muscles, improves the cardiovascular system, pumps oxygen to the brain, keeps you flexible, uses calories that would otherwise go to unnecessary weight gain, and helps to give you a better quality of sleep, as well as a sharper frame of mind. Your confidence and sense of relaxation and well-being also increase dramatically. The list goes on. Don't think about it, don't discuss it, don't question it. Just do it.

The benefits of daily meditation are only now beginning to be acknowledged in the Western world, and the practice is gaining greater acceptance as current meditation methods become less connected to Hindu or Buddhist traditions.

This is an extraordinary technique that is likely to become the foundation for all disease prevention and health enhancement programs in the 21st Century. Physiologically, meditation helps you consciously slow your mental and physical activity. This allows you to achieve a deep level of relaxation over a long enough period of time, resulting in an enhanced condition of physical energy and calmness.

A growing body of scientific research over the past quarter-century into the short- and long-term effects of meditation strongly indicates that the practice significantly lowers blood pressure, decreases dependency on stimulants and depressants, and increases the ability to sleep soundly.

People who have meditated for periods of five years or longer report greater tranquillity and optimism, a significant improvement in their ability to concentrate, and a higher tolerance for unexpected events in their everyday lives.

During this period of continual change and high stress conditions, all four of these daily health habits — nutrition, sleep, exercise, and meditation — represent a crucial foundation for every other success in our lives.

strategy 14

Increase

Your Free Days

To Multiply

Productivity

The world is entering a new time zone, and one of the most difficult adjustments people must make is in their fundamental concepts and beliefs about the management of time. The time system which is ingrained in all of us is based on factory time. Composed of two kinds of time — work time and off-time — factory time was an essential element of the bureaucratic organizations which prevailed for so long. Since processing information in offices and operating huge machines in factories required people to be working all at the same time, factory time helped make these organizations efficient.

The work ethic that most of us were taught attempted to maximize the time which was considered to have the most value in this system — work time.

The amount of overtime a person put in became a measure of loyalty. Ambitious managers boasted about the number of years they'd gone without a vacation. Even the language that evolved conveyed the inherent value judgment we held of non-work time — it was time "off."

But the emergence of knowledge and innovation as the world's most valuable resources is changing our concept of time. The integration of microchip technology in a global, 24-hour business world, is rapidly making factory time obsolete. Where results are concerned, time no longer equals money. The trouble is, most people, even many entrepreneurs, are still punching an imaginary clock.

The Entrepreneurial Time System™

BUFFER DAYS
In-between days that allow us to do all the things that set up rejuvenating Free Days and productive Focus Days

FREE DAYS
Days that are totally free of all activities related to business, their purpose is rejuvenation

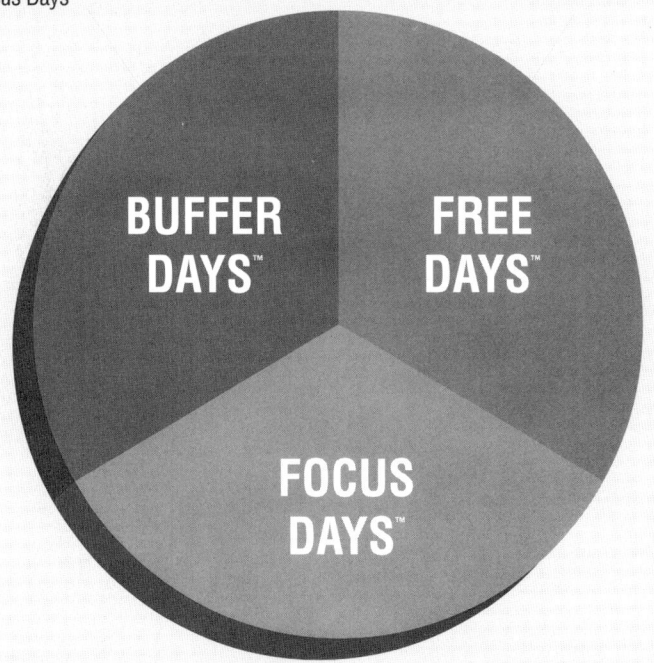

FOCUS DAYS
Days scheduled exclusively for the purpose of creating financial results and opportunity

What we need is a new concept of time which allows us to structure our lives to make optimum creative use of our lives. There is a new system of time management — designed for entrepreneurs — that has been found to increase both quality of life and productivity. It's based on three kinds of days:

Free Days, Focus Days, and Buffer Days.

Free Days are days that are totally free of activities directly related to making money. Free Days rejuvenate us physically, mentally, and spiritually. They allow us to put our lives into perspective and lay the foundation for long-range planning. Free Days enhance creativity, which is essentially the most valuable thing we have to offer in the marketplace.

Focus Days are days scheduled exclusively for the purpose of creating financial opportunity and results. Most people allow themselves to be sidetracked continually by activities which are neither productive nor rejuvenating, and consequently they are never able to benefit from the enormous opportunities which present themselves every day.

By focusing our energies on only those activities which produce the greatest short- and long-range results, we gain tremendous energy and satisfaction.

Buffer Days are in-between days. They give us the time we need to handle all those things which would otherwise drain our energy and focus, and enable us to set up excellent Focus Days and Free Days. Buffer Day activities might include cleaning up messes in our lives (both figuratively and literally), delegating job functions, or developing new personal and professional skills.

The simplicity of this system belies its tremendous advantage over the traditional way of viewing time. In order to be creative and innovative, rejuvenation is essential. So unlike a reward for hard work in the old system, taking Free Days today is actually a necessary precondition for achieving success and optimum productivity.

In fact, the more free time that entrepreneurs take, the better their results.

And the better their results, the more free time they can take. If you've already made the two entrepreneurial decisions, this may seem like a hard concept to accept, even if you don't admit to being a workaholic. But as an entrepreneur, you're in the best position to organize your time in the most enjoyable, satisfying — and productive — way. Very often, the most productive time is the period of ten days immediately following a stretch of Free Days. You return to work with a new perspective, a higher energy level, and very probably a breakthrough.

Take one vacation a year, get one breakthrough. Take two vacations a year, get two breakthroughs. Take three, get three. Plan at least five breakthroughs a year, and preferably ten.

s t r a t e g y   15

Visit Mountains,

Forests, Deserts,

And Oceans

We are biological, not technological, creatures. In the midst of the microchip revolution, it's important to remember that our fundamental connection is with nature. When we become too enmeshed in the complexities and possibilities of advanced technology, we lose the sense of who we are.

Too much machinery and electronics cuts us off from the restorative power of natural rhythms and cycles. After awhile, life begins to feel, well, unnatural.

People who visit natural environments can't help but be overwhelmed by the strength of oceans, the vastness of deserts, the tranquillity of forests, and the magnitude of mountains. It is here that they feel closest to God — whatever their definition or image of God may be. You never hear someone say they're logging on to an electronic bulletin board to feel closer to God. Technology may be extraordinarily useful, but it's seldom spiritually transforming.

Unlike computer technology, which is extremely complex and difficult to understand, nature is even more intricate and complex — yet it always makes perfect sense. If you just sit quietly and listen for awhile, nature explains itself clearly and completely.

Natural environments — especially mountains, forests, deserts, and oceans — have a soothing, therapeutic, and rejuvenating effect on people who are caught up in technological civilization.

We regain perspective and we develop a bit of humility. The cleaner air also helps us sleep better and improves our appetites. In the technological environment, things are seldom the same from one year to the next. But in nature, essential features and forces remain the same for centuries. When we visit a natural site that is the same today as when we visited it 30 years ago, we are reminded that nature is the only system that has stood the test of time since the beginning of time.

The mountains, forests, deserts, and oceans were there long before the first words were spoken by human beings — five thousand generations ago.

A veteran mountain climber describes it: "I like the mountains because they don't give a damn. They just sit there. When I go out climbing, I leave everything civilized behind me and get back in touch with a deeper life."

The more our civilization moves forward on the tidal wave of technological progress, the more we must make a special effort to stay in touch with natural environments. If we let the natural environments — mountains, forests, deserts, and oceans — do their work, we will always be nourished and revitalized. We will come back wiser and better people. That is simply the way we are. We can't help it. It's our nature.

Nature stands the test of time

s t r a t e g y   16

Decide How Long You

Are Going To Live,

And Plan Backwards

before reading this chapter, answer this question: "What age will you be when you die?" The fact is, most people don't think they know how long they're going to live, but when pressed, they generally do have an age in mind, and they answer this question surprisingly quickly. And with absolute conviction. If you have an age of 75 in mind, you will construct your life in a far different way than if your age is 100.

In fact, everything that we do tends to make sure that this subconscious "deadline" becomes a self-fulfilling prophecy.

Our health habits, our financial plans, our friendships, and our sense of purpose are all designed in such a way to get us to a specific age in life, and no further.

So now ask yourself how you arrived at your estimate for your longevity. Again, if you're like most people, your reasons are based on your family history, actuarial tables, or a close and powerful role model in your life. And of course you'll be right. (Don't read on unless you're prepared to accept the possibility that you may be wrong.)

Today, most doctors agree that current medical technology can increase the number of years that you enjoy a reasonable quality of life. This alone has dramatic implications for how you plan your activities for the next 20 or more years. In addition, there are several emerging factors in the technological and scientific

When exactly do you plan to die?

POSSIBLE REVISED DEADLINE

The possibility of an extended lifetime

Possible breakthroughs in medical science

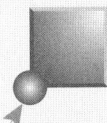

THE AGE WHEN YOU PLAN TO DIE

How do you want to be the year before?
- Physical
- Mental
- Financial
- Relationships
- Self-Assessment

PSYCHOLOGICAL REASONS WHY PEOPLE DIE

- No friends
- No money
- No purpose

TODAY

THE NEXT 3 YEARS

Based on how you want your life to be the year before you die — what do you need to focus on over the next three years?

BIRTH

world which may actually help you live longer, much longer, than previously thought possible.

It is conceivable that within the next two decades, major breakthroughs will take place in retarding or reversing the aging process in our cellular structure.

Genetic engineering and body replacements are two areas that show the most promise. Medical scientists are already working on ways of introducing microscopic robots into the bloodstream to repair all structures of our bodies on a continual basis. At the same time, the application of microtechnology to prosthetics will produce artificial replacements for many organs in our bodies, both external and internal. This is already happening to a far greater degree than most of us realize.

Unfortunately, many people stop living for psychological or financial reasons rather than medical reasons. They either run out of money, they run out of friends, or they run out of purpose. No matter how healthy you are, if it looks like you are going to spend the rest of your life in poverty, you won't have much desire for staying around. If all of your closest human relationships are gone, your future will seem very bleak. And if you run out of purpose, you also run out of anything to look forward to — except perhaps the end.

Now, let's reverse our thinking on this subject: If you could choose to live much longer than what is currently accepted as reasonable — say to 150 — how much money would you need? How could you ensure that you had close, warm supportive friendships until that age? And what kind of purpose would keep you excited and motivated over all the years between now and then?

These are not trivial questions. Your answers will have an immediate impact on your notion of how long you actually want to live. Your answers will also transform your present attitude toward your physical health, your career, and your family. And it will transform your perspective of the turbulent times we're living through — because you know you'll be around long after the microchip revolution has totally transformed the present structure of society.

strategy 17

Plan Your Life

In Three-Year

Quantum Leaps

We live in the age of the quantum leap. The term "quantum leap" comes from physics, but in everyday life it means a "sudden, dramatic improvement." During this great crossover in human society caused by the microchip, we are seeing dramatic improvements every day in virtually every area of technological endeavor. The term "quantum leap," like "breakthrough," may sound tired, but it is nevertheless accurate. Anything less would belittle the magnitude of the changes currently taking place. The big question is: How can we keep up? Certainly, none of our traditional education and training has prepared us for this environment of accelerating change.

What is needed now is a new kind of personal "school" that is uniquely geared to each person's future — so that he or she can continually make adjustments and improvements to the "curriculum."

Such a school would consist of a planning structure and methodology that will help each of us — as unique individuals — to maintain control over the pace of change in our lives, so that we can enjoy our growth and progress rather than feel victimized by external changes. Such a school would be based on achieving personal "quantum leaps" that keep pace with the quantum leaps that are occurring everywhere around us. There are already many individuals — confident entrepreneurs — who are living their lives in this quantum-leap fashion, even though they may not be conscious that they are doing so. Every three years, they demonstrate a "sudden, dramatic improvement." But their progress in personal and business life doesn't

Structure of a Quantum Leap Lifetime

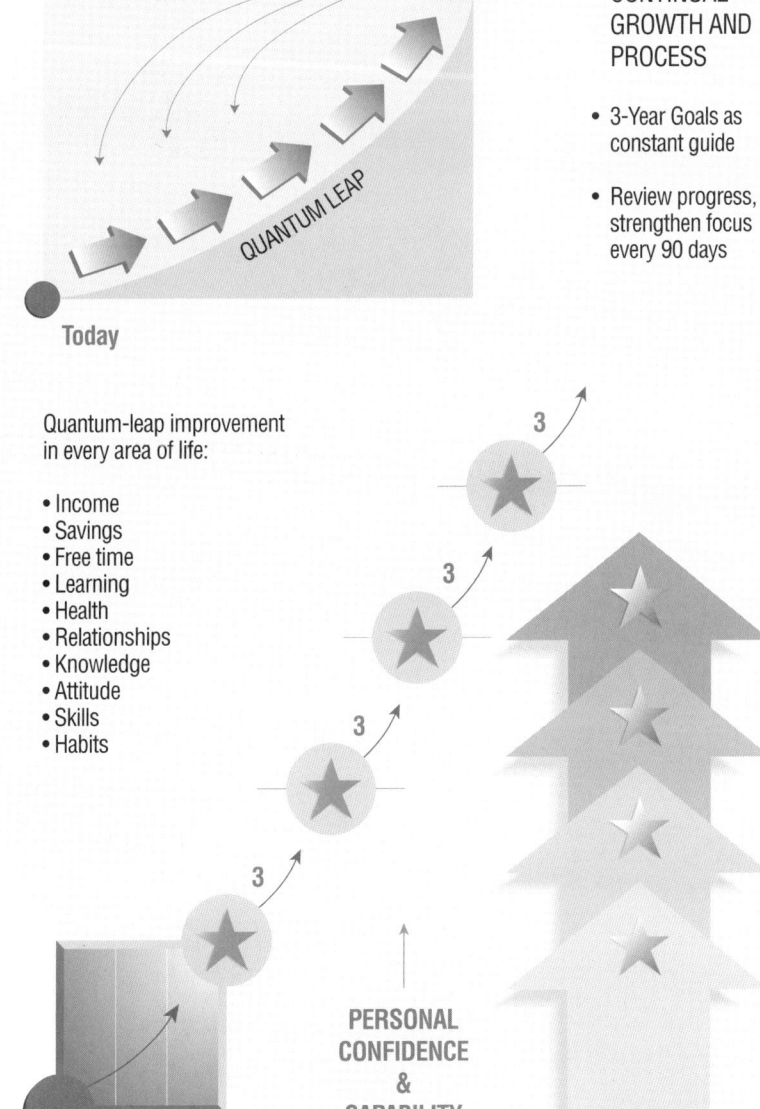

CONTINUAL GROWTH AND PROCESS

- 3-Year Goals as constant guide
- Review progress, strengthen focus every 90 days

Quantum-leap improvement in every area of life:

- Income
- Savings
- Free time
- Learning
- Health
- Relationships
- Knowledge
- Attitude
- Skills
- Habits

PERSONAL CONFIDENCE & CAPABILITY

happen by chance. They plan on it. They visualize it. They build the structure to support it. And they take the necessary steps to bring it into reality. Often, the quantum leap is made possible by taking advantage of new opportunities created by technology, but it can also be the result of a new strategic affiliation, a better support staff, or a tight focus on a key area of expertise, or all of the above.

Three years is an effective time frame for personal planning, goal-setting, and personal improvement. It's long enough to imagine a different scenario — free of current problems, obstacles, deficiencies, and frustrations — yet short enough to work out the specific steps you need to take to achieve the results you're after. If you close your eyes right now and imagine where you want to be three years from now, you'll come up with some very vivid pictures of big improvements.

The quantum leaps you visualize may be in the areas of increased income, greater savings, more free time, new capabilities, better delegation, bigger markets, improved personal relationships, better health, community involvement, or a combination of all of these.

If you then write down your goals and attach deadlines for their achievement, your mind will begin to "buy into" the reality of the quantum leap. Almost immediately, you will find yourself making decisions and taking actions that lead to the accomplishment of these goals.

Planning your quantum leaps in periods of three years also gives you the time you need to change your habits.

This is especially important since most personal improvements in life invariably come down to a change in habits, which takes consciousness, commitment, focused effort, and time. There are some things we can change quickly in life, but not our overall structure of personal habits. Knowledge can change in a few seconds, attitudes can be shifted in minutes, and learning new skills may take a few weeks or months. But a conscious effort to change fundamental habits takes about three years. If all you did was visualize your quantum leap and then never thought about it again, chances are it would never materialize. That's why it is important to divide the three-year time frame into smaller, more manageable periods to implement your decisions, actions, and practical improvements. Ninety days work very well. This period gives you an opportunity to make noticeable progress toward your three-year goals and can be broken down into weekly units for you to achieve an even greater sense of continual improvement. Throughout the three years of making decisions and taking actions in this structure of twelve 90-day jumps, you will notice an acceleration of growth, learning, and achievement. By the end of this "quantum leap school," there will be a dramatic improvement in all the areas of personal and business life where you have set greater goals for yourself.

Once the structure of the quantum leap is mastered, it can be repeated at a higher level of achievement. This process of making dramatic overall improvements every three years can become a fundamental approach to constant growth for the rest of your life. Let's say you're 40 years old now, and you plan to live until 85 — another 45 years. That gives you the possibility of 15 three-year quantum leaps.

Every three years, your sense of achievement, capability, and confidence increases.

Every three years, your quality of life improves, your relationships become more creative and positive, and you feel even more certain that the following three years will be even better. Entrepreneurs who consciously plan for quantum leaps lead remarkably different lives from the majority of people today who are still on the defensive with regard to technological change.

Entrepreneurs who plan for quantum leaps feel more in charge of themselves and their circumstances.

Regardless of what is happening to other people, these quantum-leap entrepreneurs are always going in the direction they want to go, learning what they need to know, and improving where they need to be better. As a result, they achieve the specific results they want out of life — regardless of what other people are doing with their lives.

Entrepreneurs who plan for quantum leaps do not fear change, nor do they see technology as a threat.

They understand that the turmoil, confusion, and disruption

being caused by microtechnology is just part of the crossover to the next stage of human development. Sooner or later, most people on the planet are going to make the transition to using electronic language as a normal part of daily life — in the same way that human beings in previous times adapted to spoken language, written language, and printed language.

Entrepreneurs who plan for quantum leaps are neither in love with microtechnology nor overawed by its possibilities.

They use many different kinds of resources to achieve their quantum-leap goals, and microtechnology is just one of them. They are not seduced by all of the advertising claims, promises, and predictions coming from the marketers of technological products and services. All of these new microchip tools, as potentially useful as they may be, are just that — tools — with no special miraculous qualities. The miracle of the microchip is not in the tools themselves, but in the creative visions and capabilities of entrepreneurial individuals who use the tools to achieve their quantum-leap improvements.

Every year we are going to continue to see dramatic breakthroughs in microtechnology. That is the essential nature of this technology. It evolves in quantum leaps, so we might as well get used to it as a normal part of life in the 21st Century. The trick to planning, organizing, and living our everyday lives in a world characterized by quantum-leap change is to create our own quantum leaps.

s t r a t e g y   18

Team Up And Prosper With Female Entrepreneurs

*f*or women, this is an age of economic empowerment and unlimited opportunity — especially as entrepreneurs. Over the past ten years in North America, the majority of new business starts have been companies owned by female entrepreneurs. In 1992, businesses owned by women employed more people — 11 million plus— than the Fortune 500 corporations for that year. This is an extraordinary trend, and again the microchip is a major cause.

Microtechnology is making it possible for millions of women to bypass the traditional bureaucratic and corporate path to employment and a business career. Ambitious women with business dreams and capabilities are staying away from the traditional business schools and the corporate ladder. The relative low cost of microchip-based tools enables a woman with an entrepreneurial concept to begin her business and grow it successfully with very little start-up capital. The availability of software programs and information bases in all areas of expertise provides a woman entrepreneur with crucial knowledge and methods that previously were the exclusive preserve of large bureaucratic organizations.

But there are deeper reasons for the emergence of female entrepreneurism: Women's attitudes and capabilities are in alignment with the growth of microtechnology.

Male values and structures are deeply imbedded in hierarchy, conformity, obedience, and internal competition. Women, on the other hand, long excluded from bureaucratic power and

Female abilities and the microchip

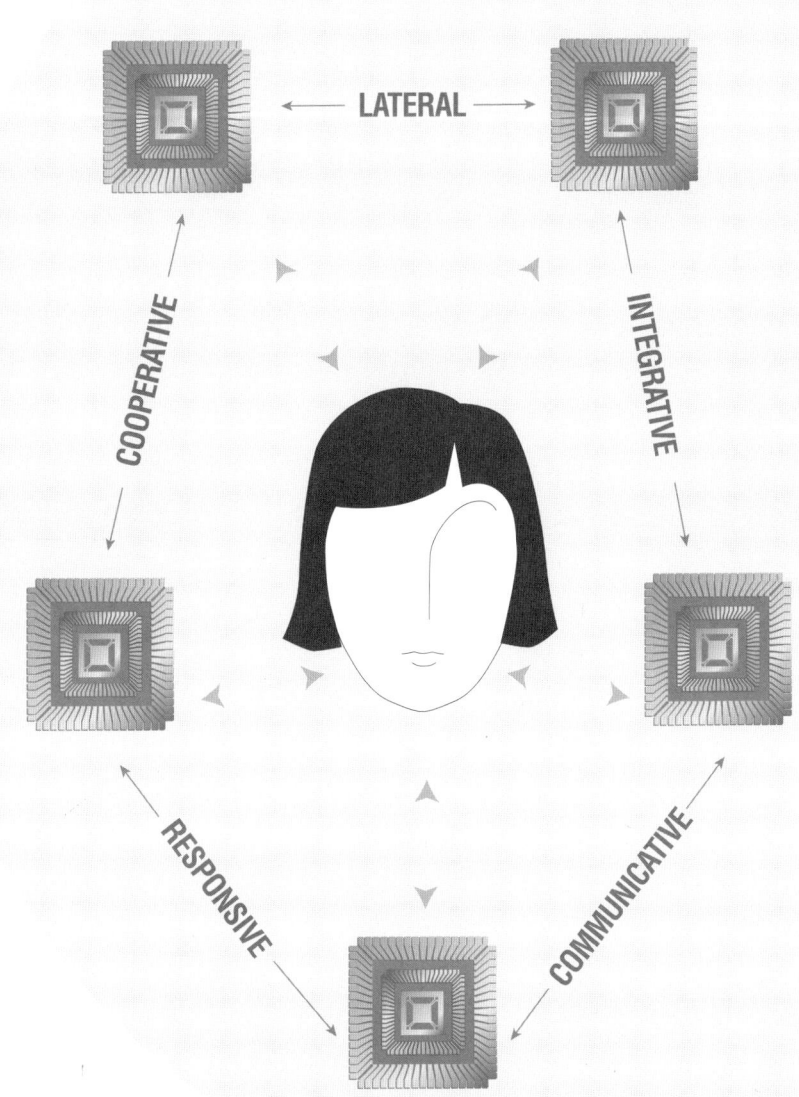

status, have learned to operate in lateral, cooperative, and keenly responsive ways. This is the way that microchip technology operates and expands — laterally, cooperatively, and keenly responsive to feedback.

We're moving rapidly from centralized mainframes to decentralized networks of personal computers. Everywhere, there is a movement away from the centralized control of decision-making and information that has been characteristic of male-dominated organizations for millennia. As logic, rationality, uniformity, and conformity — bureaucratic qualities — are increasingly programmed into the microchips and software, human beings are freed up to be human — that is, intuitive, creative, supportive, and compassionate.

Women are skilled and confident in this new environment. Their roles as mothers — or having their mothers as role models — give female entrepreneurs an instinctual advantage as communicators, integrators, facilitators, counselors, catalysts, problem-solvers, and nurturers. They've learned from childhood how to provide a wide range of services to others within the family structure; now they are expanding these into the marketplace.

Strange as it may seem, the disadvantage of being a woman through much of history — ignored as leaders, suppressed as self-expressive individuals, denied formal education, discounted as creators, forced to do menial work, kept out of power — is turning out to be a big entrepreneurial advantage for the future.

Thousands of years of training as political, economic, and cultural subordinates have suddenly, at the beginning of the 21st Century, provided women with a psychological and practical advantage to integrate, facilitate, and nurture the growth of a global electronic economy. As far as we know, this is the first time in history that we have had an economic restructuring that not only allows the full participation of women, it absolutely depends upon it.

Ironically, it is now bureaucratically-trained men who are facing the greatest difficulty. The worst casualties of the microchip revolution so far are those men who cannot function outside of the hierarchical chain of command. Having focused on conformity to superiors within a rigid structure all their lives, these men now find it impossible to create their own opportunities. They cannot adapt to an economic environment characterized by free-flowing creativity, communication, and constant change.

While traditional male-dominated leadership continues to resist change through bureaucratic methods and structures, female entrepreneurs are creating millions of smaller power centers that speed up the process of change. As a result, the most powerful and creative entrepreneurial networks at the turn of the 21st Century will be those with the largest participation of women entrepreneurs.

A key to success, therefore, for all entrepreneurs — both men and women — is to establish strategic alliances with companies owned and operated by women with big ambitions.

strategy 19

Ask Everybody You Meet The Three-Year Question

*t*he good news is that people have more choice than ever before in history about their direction in life. The bad news is that most people have no idea in which direction they want to go. We can no longer depend on our traditional leaders to provide leadership. In fact, it's clear that governments, the military, clergy, teachers, and parents are having trouble finding direction themselves. In many cases, the authority figures we used to rely on have lost credibility. While many aspects of traditional society, based on hierarchical authority, are weakening, people are left with the job of making sense out of their own futures.

As a result, it's increasingly important for us to become self-directed — establishing and achieving flexible personal goals regardless of the changes that are taking place externally.

Most people have images in their minds about how they would like their future to unfold, but these images tend to be vague and incoherent. We can actually help each other articulate and organize them into powerful goals simply by asking the question:

"What has to happen over the next three years for you to feel happy about your progress, both personally and professionally?"

Three years is a good time frame — long enough to allow for significant results; short enough to visualize realistically. This question serves six useful purposes:

The question that creates futures

"If we were meeting here three years from today, and you were looking back over those three years to today, what has to have happened over those three years, both personally and professionally, for you to feel happy with your progress?"

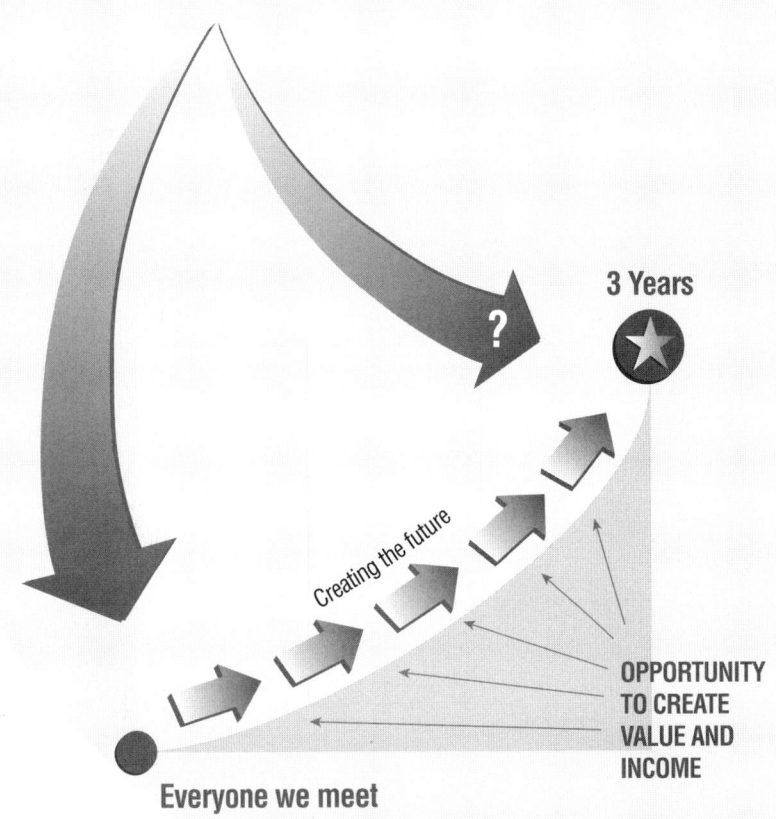

1. **It immediately indicates whether another person's fundamental commitment is to the future or to the past.**
   Some people — out of fear of what may lie ahead — try to live their lives based on past experiences and accomplishments, and it is extremely difficult for them to think in terms of something new and different. They lack the courage to dream. Since it is virtually impossible for you to involve these individuals in future-oriented possibilities and projects, it's important to avoid wasting time with them right from the beginning.

2. **A person's willingness to answer this question immediately indicates whether or not she or he trusts you.**
   Without trust, there is no basis for a relationship.

3. **Asking such a direct question elicits a more honest answer, and helps other people to be more creative and strategic.**
   Questions are powerful tools. Being asked a question forces us to think about things in new ways, and at a more fundamental level. You can change a person's concept of the future — in positive and creative ways — simply by asking this question and listening to the answer.

4. **When someone else tells us what his or her most important goals are, it puts us in a much better position to assist that individual, either through**

direct action, or by referring him or her to other useful people and resources.

In a very real sense, this kind of question is at the core of all new products and services in the marketplace.

5. **The three-year question leads to fascinating conversations regarding the most important subjects.**

   Many people have learned how to converse without saying anything that is important or meaningful. Generalities, pleasantries, and chit-chat are just passing words. But when people talk about their future — the one to which they are truly committed — they talk about very specific actions and accomplishments that require creativity, commitment, and courage, and this is always inspiring to hear about.

6. **Helping other people to think about their future helps us to be clear about our own.**

   If we ask this question often, we will become aware that many people around us are striving to improve their lives, and this gives us the courage to make decisions and take action. Helping others become clearer about their future and excited about their prospects is perhaps the greatest service we can offer to others, but it's not simply an altruistic exercise. Some answers often open the door for business opportunities and joint ventures. Surrounding yourself with people who are equally motivated and enthusiastic about the future creates a mutually supportive environment that helps you achieve your own goals.

strategy 20

Grow To The Genius

Level Of Personal

Confidence

*g*enius is possible in all areas of human experience. Most people, however, when they hear the word "genius," automatically think of great musicians, artists, inventors, or scientists. But genius is a function of extraordinary self-confidence, and this can occur at any time, in any place, in any area of human capability.

Every period of history has its own special area of genius. At the beginning of the 21st Century — in the age of the microchip — it seems to be emerging most of all in the area of entrepreneurism, among those men and women who have decided to be economically self-reliant.

Not every entrepreneur is a genius, however, because many never become self-confident in any area except that of making money. The true entrepreneurial geniuses are those who have extended their economic self-confidence into every other area of their lives.

There are seven specific levels of self-confidence that entrepreneurs need to master to allow their unique genius to develop over a period of many years:

1. **Confidence about health**
   A commitment to developing one's unique vision and abilities in life requires a lot of energy and stamina. Also, we want as much healthy time on the planet as possible to translate our self-confidence into action and achievement.

The seven levels of personal confidence

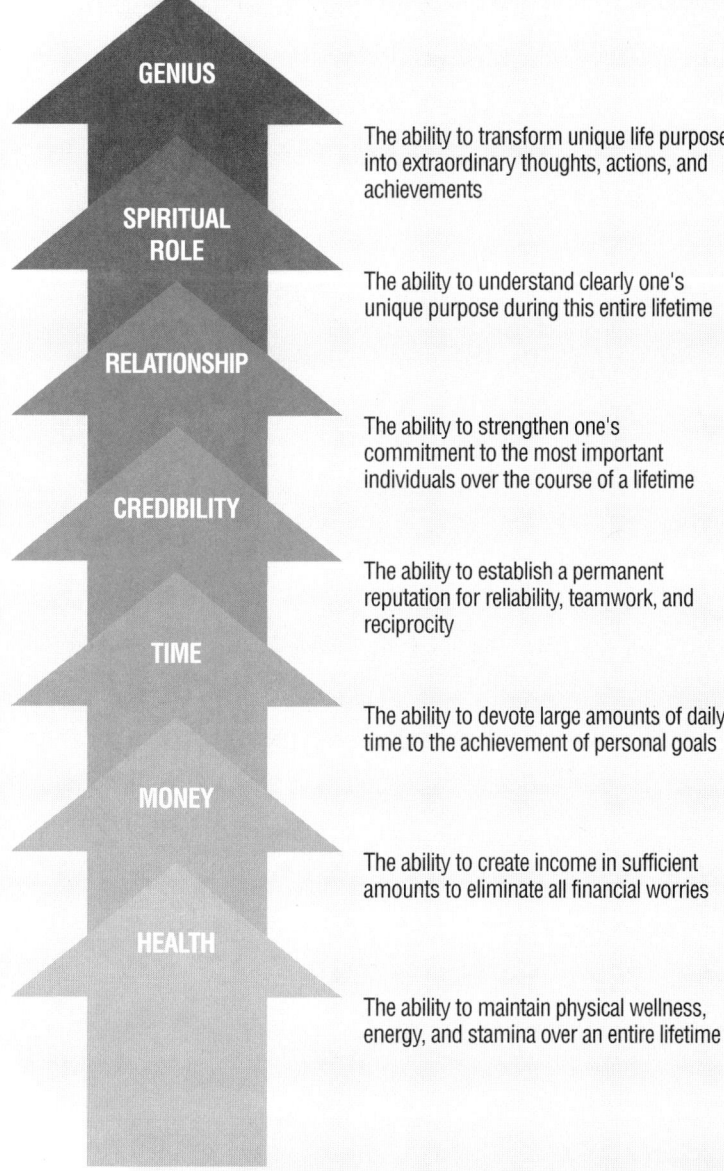

A lifetime structure for creating the conditions of entrepreneurial genius. Each of the seven levels of confidence strengthens the other six.

2. Confidence about money

You can't do much if you don't have the dollars. This means having an independent way of earning income so that others cannot control one's economic activities and results. The two entrepreneurial decisions (Strategy 1) and getting away from bureaucratic organizations (Strategy 2) are crucial to any growth of confidence in this area.

3. Confidence about time

To control our activities in life, we have to control our time — time for making money, time for relationships, time for improving our health, time for improving our knowledge and capabilities, time for enjoying ourselves, and time for understanding our unique contribution to others. In today's world, controlling one's time costs a lot of money. That's why confidence about money (Level 2, above) is so important.

4. Confidence about credibility

No one gets taken seriously in this world unless he or she has credibility — not credibility about brilliant ideas, or heroic deeds, but credibility about daily habits and performance.

There are four crucial "credibility habits:"

Showing up on time.
Doing what you say.
Finishing what you start.
Saying please and thank you.

These simple habits may seem self-evident, but the failure to observe them is probably the biggest cause of loss of credibility in our relationships with others.

5. Confidence about relationships

If you look at the first four levels of confidence, it's easy to understand why so many people have troubled and unsatisfactory relationships. Many people do not have the energy, money, time, and credibility necessary to protect and nurture their most important relationships. Once people achieve confidence in their closest relationships, it becomes possible to establish and maintain essential, mutually beneficial associations with others who are also developing their genius.

6. Confidence about spiritual role

"Spiritual," here, has nothing to do with religion, or, if so, only coincidentally. Confidence about one's spiritual role in life means that a person understands very clearly why he or she is on the planet. In other words, what is the unique creation of thought and action that is going to be developed over the course of one's lifetime?

The answer to this question does not come from any sacred book, but rather through constant creative exploration and experimentation over a long period of time. Once people answer the question of spiritual role for themselves, it becomes apparent to others. They radiate a calmness, a certainty, a sense of simplicity, and a sense of humor and humanity.

7. **Confidence about genius**

    Once the spiritual role becomes clear, the development of genius becomes a matter of concentrated effort over a long period of time. From the moment that a person develops confidence about his or her unique contribution, it takes about ten years for his or her thoughts and actions to take on genius-like characteristics.

The age of the microchip is really the beginning of a period in human history when vast numbers of human beings can become geniuses — following the path of increased entrepreneurial confidence throughout a lifetime.

Until recently, most people's time, energy, and concentration had been used up in the activities of physical survival. But microtechnology changes all of that — it frees us up from physical labor and struggle on a massive scale.

The current view is that the elimination of human labor is a terrible threat to society, and that all of the organizing structures of society will collapse.

But there is another — entrepreneurial — way of looking at this:
All human development to this point in history has been simply to get us to the point where our entire lives — supported at all points by microtechnology — can be devoted to developing entrepreneurial genius.

With microtechnology creating unlimited economic opportunities for entrepreneurs, there is now nothing to stop us from achieving total self-confidence in every area of our lives.

background

The

Srategic Coach

Program

The concepts and strategies in this book have been developed over the past ten years within the context of The Strategic Coach Program. The Program is a three-year focusing school for experienced and successful entrepreneurs, with participants from throughout North America and abroad.

The Strategic Coach is designed for three purposes:

1. To eliminate from an entrepreneur's life all the "stuff" and "messes" that interfere with productivity.
2. To increase the entrepreneur's concentration on the most important money-making relationships and centers of influence.
3. To increase dramatically the amount of free time that can be devoted to personal relationships, increased health, and so forth — all of the personal interests that make being an entrepreneur worthwhile.

The Strategic Coach is a system.

The Program consists of 12 fundamental "breakthrough strategies," mastered over a period of three years, that systematically simplify and clarify everything that the entrepreneur thinks, says, plans, does.

These 12 strategies, when implemented and integrated with each other, create a permanent condition of high energy and creativity in which all of the entrepreneur's resources are focused entirely on the most important opportunities.

The Strategic Coach is a philosophy, a structure, a process, and a methodology.

The Program lasts three years, which is sufficient time to establish and change fundamental entrepreneurial habits. Over that period there are 12 one-day workshops at 90-day intervals, each of which enables entrepreneurs, in the company of other entrepreneurs, to achieve a deeper understanding of the 12 focusing strategies.

Entrepreneurs who enroll in The Strategic Coach Program have come to a point in their careers when all the skills, experience, and opportunities are there, and it's time to put everything together.

What is needed for further progress – for their next quantum leap – is a clear-cut vision, a structure and a strategic gameplan to keep them on track, a group of other high performers to keep them honest, and an experienced coach to keep them focused.

If you would like further information about The Strategic Coach Program, or other Strategic Coach services and products, please telephone 416•531•7399 or 1•800•387•3206. Or visit our website at *www.strategiccoach.com*.